Praise for *Restoring Our Sanity Online*

Three and a half decades ago, when I invented the web, the intention was to allow for collaboration, foster compassion, and generate creativity. It was a tool to empower humanity. Yet in the past decade, instead of embodying these values, a few social media platforms have played a part in eroding them. *Restoring Our Sanity Online* reveals the web's fascinating journey with powerful anecdotes. The book's call to action for a human-centered web aligns with my original vision. Mark Weinstein presents a new paradigm for the web and social media that empowers users, reshapes data storage around user-controlled Solid Pods, protects democracy, and eliminates surveillance capitalism. Its vision, which I share, is a digital future that prioritizes human well-being. This is a vital read.

—Sir Tim Berners-Lee
Inventor of the Web

Technology has a purpose: to do good and to share. *Restoring Our Sanity Online* is a timely and essential read for those seeking an approach to social media that aligns with this. Mark Weinstein offers a refreshing vision—users are customers to serve, not products to sell. The book presents practical solutions for social media that doesn't rely on tracking users or targeting them with ads. It provides a blueprint for sharing profits with users while championing privacy, mental health, and authentic connections. This is a must-read for anyone concerned with where we are today and looking for a better path forward.

—Steve Wozniak
Co-founder, Apple Inc.

Mark Weinstein has long been a leading voice advocating for the responsible use of social media. In this important book, he brings decades of wisdom to bear on one of the most consequential issues of our times—how we can deploy the extraordinary power of social media and AI in ways that align with our collective well-being over the long term. This is essential reading for business leaders, policy makers, and citizens at large.

—Raj Sisodia
Co-founder, Conscious Capitalism Inc.

Mark Weinstein's *Restoring Our Sanity Online* is a compelling journey into the heart of the digital age, where the promises of social media collided with the stark realities of its influence on our lives. Combining personal anecdotes and a critical analysis of Big Tech's evolution, Weinstein offers a hopeful narrative on how we reclaim the internet for its intended purpose–connection and community. This book is a must-read for anyone looking to understand the impact of social media and seeking practical solutions to create a healthier online environment.

—Dr. Marshall Goldsmith
Bestselling author, *The Earned Life; Triggers; What Got You Here Won't Get You There*

RESTORING OUR SANITY ONLINE

MARK WEINSTEIN

RESTORING OUR SANITY ONLINE

A REVOLUTIONARY SOCIAL FRAMEWORK

WILEY

Published by John Wiley & Sons, Inc., Hoboken, New Jersey.
Published simultaneously in Canada.

For general information on our other products and services or for technical support, please contact our Customer Care Department within the United States at (800) 762-2974, outside the United States at (317) 572-3993 or fax (317) 572-4002.

Wiley also publishes its books in a variety of electronic formats. Some content that appears in print may not be available in electronic formats. For more information about Wiley products, visit our web site at www.wiley.com.

Library of Congress Cataloging-in-Publication Data:

ISBN 9781394273966 (cloth)
ISBN 9781394273973 (epub)
ISBN 9781394273980 (epDF)

Cover Design: Wiley
Cover Images: © 7AM/Adobe Stock,
© Fotograf/Adobe Stock
Author Photo: Courtesy of Jonathan Young

SKY10081180_080624

With love to my dad and my kids:

You always inspire me

Contents

Introduction

I AM A passionate fan of social networking. I love its dazzling tech at our fingertips allowing us to share our lives anywhere, anytime. As one of the earliest founders in this remarkable communication revolution, I found it exhilarating to participate in the movement bringing it to life.

But my love comes with a caveat. The original vision and execution of social networking were based on an enchanting premise. Geographic distances and technology limits were eliminated. We stayed connected with friends and family in exciting and enriching new ways. An added bonus was the ability to expand our friend circles, join like-minded communities, and explore novel interests.

In its brilliant ascent, social media was leading us toward a more compassionate and connected world. Uh-oh—did we take a wrong turn somewhere?

Today, Big Tech is driving us, our kids, and society mad. The current genre of social media is a far cry from its birthright. We have a crisis on our hands, and most of us know it. We feel small and unprotected. Tech giants rule with impunity. We're trapped in a distorted landscape, pawns of their overt and nefarious manipulations. Here's a sobering stat: A Pew Research survey in 2024 found that "about two-thirds of Americans, 64%, said they think social media is bad for democracy."[1]

It's easy to think we're in uncharted territory. But the perpetrations of Big Tech—manipulations, monopolies, and profits at all costs—aren't necessarily new. There are striking parallels in Big Agriculture and Big Energy, and valuable lessons to glean.

Get ready for a fascinating read that will keep you jaw-dropped, incredulous, and chuckling along with the book's reveals.

Together, we'll embark on a journey to discover an authentic, uplifting, and achievable transformation of social media and Big Tech. This is the game plan for an epic reboot. What would social media look like if it nourished our critical thinking, mental health, privacy, civil discourse, and democracy? Is that even possible? Stay tuned.

This book is for all of us—from the curious and uninitiated, to casual and heavy users of social media, to parents and teachers, to techies and entrepreneurs, to investors and elected officials. Everyone is affected by Big Tech, and everyone is part of the solution.

It's time for velocity. What we need is insight, understanding, and an implementable plan to catapult forward. How do we get there? By the end of this book, you'll know the answers.

In your hands is the framework.

1

My Adventures in Web1 and Web2: The Rise of Surveillance Capitalism

I WAS EARLY, but I wasn't the first. Social media was born in the late 1990s. A handful of Web entrepreneurs were tinkering, independent of each other. It was a classic hundredth-monkey effect.

In 1997, my family gathered from many points for a vacation in Stanley, Idaho—population, 110. It's a beautiful place. The landscape near the Sawtooth Mountains is idyllic both for being together and for solitary enlightenment. Perfect for expansive thinking.

Hiking around pristine Redfish Lake, the conversation was completely unintended. My 10-year-old nephew, Justin, and I fell far behind our family members. We were chattering about using the still-newfangled Web to keep us all together. Over 25 years ago, email was already antiquated and notable for its visceral void. Web-based communication tech was in its infancy. It started like this: "Uncle Markie, wouldn't it be great if we could . . ."

Ten minutes into our conversation, I was overtaken by a tingling sensation, an awareness. The words jumped off my tongue. "Justin, I am going to start a company to do this, and I'm going to give you 10% of whatever I own."

Game on.

At dinner that night with our extended family, we imagineered, taking notes on a couple of napkins about what we wanted to share online: photo albums, chats, discussions, address books, recipes, a family newspaper, a calendar, wish lists, birthday reminders, and more.

This was Web1, as it is now known. There was plenty of capital available. The internet horse was out of the barn and everyone wanted to hitch a ride—designers, engineers, marketers, executives, you name it. Web1 had its own singularity—everyone was focused on serving the end users. No one was yet enamored with their data, or algorithmic manipulation of the ads they were served, or with their newsfeeds and purchases. Eyeballs were the valued commodity; revenue models and monetization would follow later. Web1 was labeled the "New Economy."

It was classic entrepreneurialism. In those days, there was no LinkedIn or Indeed—I placed ads in the *Albuquerque Journal*'s help wanted section and interviewed candidates at my kitchen table. My 1,445-square-foot home became company headquarters. Today, software engineers are everywhere in the world. At that time, there were hardly any engineers or graphic designers experienced with Web applications. Salt that with the way people accessed the Web, via dial-up. Efficiency in programming was paramount, as most sites loaded slower than pouring molasses, and many not at all. The first graphic designer I hired was well established and highly regarded. His designs were beautiful. But his Web pages wouldn't load.

In a stroke of luck, the State of New Mexico gave my just-birthed company a $300,000 grant to remain in the state. At the hearing to consider my grant request, they first denied it—then the chair, Mr. Garcia, pulled the committee into a back room. Moments later, they reemerged.

"Mr. Weinstein," Mr. Garcia began, "several years ago, we rejected Bill Gates's application and he left the state . . ."

Gates's dad, I later learned, beckoned him to come home to Seattle for parental funding.

Mr. Garcia continued, "After reconsideration, we are approving your request for a $300,000 In-Plant Training Grant."

Wow! Thank you, Mr. Gates. The state also provided and subsidized sweet offices in Albuquerque's newly built Science and Technology Park and added a custom-built server room. We were off to the races.

We grew to about 70 team members in all. Along the way, we partnered with Sun Microsystems and Oracle to build the largest commercial installation of servers in New Mexico. Every weekend we ran a full-page print ad in *USA Today*, promoting this newfangled social Web experience. There were billboards touting our sites. In 1998, traditional marketing techniques were the best way to reach and attract users. Our flagships, Superfamily.com and SuperFriends.com, made *PC Magazine*'s "Top 100 Sites" list three years in a row.[1] It was one of the most prestigious accolades for websites at the time. We were participating in a new paradigm, "Community Portals" (later to be called "Social Networks").

Web1's colossal fall from favor with investors in 2001 hailed its curtain call. Virtually overnight, what had been an easily accessible pot of investor capital, limited only by how much dilution was palatable, dried up. Understandably, Web1's "New Economy," based on a site's number of users and their anticipated future monetization rather than the tried-and-true measurement of actual revenue, profit, and loss, fell from grace. Investors panicked before seeing their companies achieve financial success. The stampede was brutal. The sudden dearth of funding caused massive failures, though everyone had bought into the paradigm. A historical event, the dot-com bubble, burst, and I was caught in the middle.

Egg on my face, guilty as charged—as I had bought into the eyeballs-first mantra and believed it was the only way to encourage hesitant newbies to step into the Web. Paradoxically, the phoenix rising—Web2 —professed the same mantra, but with a new twist: Surveillance Capitalism. Personal data analytics and targeted marketing would resolve the revenue conundrum, manipulate those eyeballs, and make profits pop like never before.

Looking back, it's both curious and foretelling. There was a purity of purpose in Web1, the dedicated focus on serving the customer, in this case the "user," the leveraging of rapidly iterating communication technologies. What I and a handful of other entrepreneurs were creating in Web1 is now known as "personal social networking." Today, over 5 billion people participate in this paradigm.[2] In Web2, it would become one of the most successful business models in history.

Web2: The Rise of Surveillance Capitalism

"Privacy No Longer a Social Norm, Says Facebook Founder."[3] The headline in *The Guardian* blared on January 10, 2010.

My mouth opened, jaw-dropped, as I witnessed Mark Zuckerberg make this bold declaration at the Crunchie Awards in San Francisco, a ceremony then widely covered in the press. Social media was invented to serve people, not spy on them. His thought process and arrogance bewildered me. His wallet was more important than the fundamental human right to privacy?

Today and for most of the 21st century, we've been participating in the greatest socioeconomic experiment in human history. It's called Surveillance Capitalism. This is the elephant in every room of our lives. Surveillance Capitalism is the modern-day business model in which everything we do—morning, day, and night—is tracked, analyzed, and monetized.

Our personal information is packaged into datasets shared and sold in a hidden ecosystem of massive data companies. These data brokers generate hundreds of billions of dollars per year by collecting our data from Web and social media companies, credit card networks, retailers, and other entities. They then provide access or sell it to whomever desires it.[4] Our data is then used to target and manipulate us—by advertisers, marketers, social media companies, politicians, governments, and every other person or entity that aims their dart directly at us. Our independent thoughts, critical thinking, and privacy are becoming relics of the past.

In the past two decades, the rise of Big Tech has placed human beings under nearly constant surveillance. Our phones, computers, Alexa, the Web, and everything we do online, everywhere we go, is all being tracked. Our relationships, our purchases, our finances, our health issues, our politics, our religious beliefs, our diets—you name it. It's estimated that, by the time we reach the age of 13, about 72 million data points on each of us have been collected.[5] This number is stunning and hard to grasp, but true. This data is monetized by commercial interests, stolen by hackers, and shared with governments.

How the heck did we get here? Courtesy of the premise and promise of Web2. In the ashes of Web1, companies like Google and Facebook arose and discovered they could reap more profits by spying on us under the guise of serving us. This solved the conundrum of Web1, and suddenly revenue flowed handsomely from their new targeting business models. Other companies quickly followed suit, spreading the Surveillance Capitalism business model across the world.

Today, these entities know more about us than our own mothers do. They know everywhere we go, every friend and family member, every search, every website, every health issue, every financial transaction, every fantasy and fear that we research. All these private moments are broadcast directly into the global data ecosystem. As Yael Eisenstat, former CIA analyst, diplomat, and Facebook employee, said in 2019, "Facebook knows you better than the CIA ever will. Facebook knows more about you than you know about yourself."[6]

So, somebody out there is collecting all your private information. Is that such a bad thing? Yes, it is. If you want to see where Surveillance Capitalism is headed, let's take a look at what is going on right now in China.

China's "social credit system" tracks its citizens and imposes punishments for what it deems to be undesirable behaviors, like playing too many video games, jaywalking, and "frivolous spending."[7] Penalties include blocking citizens from getting good jobs and preventing their kids from attending top schools. And Chinese citizens have even been thrown in jail for making jokes and sharing satirical memes[8] about the authorities in "private" social media chat groups.

With the help of AI, Surveillance Capitalist companies employ psychologists and data scientists to analyze and covertly manipulate our emotions, thoughts, votes, and purchases. This is a good career path if you enjoy snooping—they're always hiring. The companies then sell access to our eyes and newsfeeds to the highest bidder. They keep us addicted to our phones, iPads, laptops, . . . The more time we spend on their platforms, the more data they can extract from us, and the more they can invite any paying customer to take aim at us so they can win the game.

With their deep knowledge of our psyches and emotions, the social media giants even deliberately serve us polarizing content. Why? Because outrage keeps us glued to our screens.

At the time of Zuckerberg's declaration, my life was cozy and bountiful, but I couldn't ignore the nagging in my soul. This was bigger than all of us; it was about the future—ours, our kids', and the world's.

■ ■ ■

I changed my life once more, relocating in 2011 to Mountain View, California, the heart of Silicon Valley. This time, Albuquerque would not suffice for fundraising; California investors and venture capitalists (VCs) were no longer keen to get on airplanes to vet a tantalizing proposition. Their funds and countless deals were now right in their neighborhoods.

What started with focus groups in my dining room in Albuquerque and then a multiyear beta project in Sunnyvale became the Facebook alternative, MeWe. Luminaries joined MeWe's Advisory Board, including Tim Berners-Lee, the inventor of the Web;[9] Raj Sisodia, cofounder of the Conscious Capitalism movement;[10] Sherry Turkle, an MIT academic and leader in tech ethics;[11] and Steve "Woz" Wozniak, cofounder of Apple.[12]

The site and app launched in 2016, built purposefully to solve the issues perpetrated by the Surveillance Capitalists dominating social media. MeWe provided the industry's first Privacy Bill of Rights[13] for its users. While carefully protecting civil discourse, under my leadership MeWe also had no ads, no targeting, no data harvesting, no boosted content, and no newsfeed manipulation. Deploying a customer-centric freemium model, revenue came from premium features, in-app purchases, and subscriptions.

Talk about another exhilarating ride! Curiously, the promise of VC funding failed to materialize—something about having landed in Facebook's backyard with a clear message that MeWe intended to be a direct full-featured competitor. Apps that were getting funded were more discreet, and most were launched as one-trick ponies, like Snapchat, Twitter/X, WhatsApp, and Instagram. They had unique and slim features that went viral quickly, allowing their teams to raise millions while not encroaching on Facebook's turf.

This freezeout caused plenty of consternation and challenges but a simultaneous freedom from VC oversight and their likely dilution of MeWe's principles. Without large, eight-figure inflows, investment capital—millions of it—was provided by a myriad of qualified investors who understood MeWe's raison d'être.

I traveled the globe—from the USA to Asia to Europe to Latin America—to secure the company's funding. So much for my geographical location choice to avoid just that. Funds raised were consistently deployed to stay competitive with Facebook and with the constant flow of new features and fads (1:1 communication, private and open groups, disappearing content, stories, journals, custom memes, dual camera videos, live video calls, voice memos, encrypted chat, personal cloud, etc.).

During my time there, the company won numerous accolades including being named a 2016 Start-Up of the Year Finalist for

"Innovative World Technology" at SXSW; a 2019 Best Entrepreneurial Company in America by *Entrepreneur Magazine*; and a 2020 Most Innovative Social Media Company by *Fast Company*.

It became a battle of the Goliaths versus the Davids (with Goliaths consistently winning) as mainstream networks peaked and new upstarts failed. Some gained outstanding traction, notably WhatsApp and Instagram. Facebook gobbled them up among countless acquisitions. All others failed and were squashed. In the same time window, the data industry became the King Kong of it all.

To my knowledge, MeWe is one of the only apps in Web history to achieve 20 million registered users and millions of dollars in revenue without influencers, a marketing budget, or VC funds. Perhaps the numbers would have been closer to 500 million users or more if Facebook hadn't stifled the good word about MeWe.

In federal court on August 19, 2021, the United States Federal Trade Commission (FTC) filed its amended antitrust complaint,[14] stating that Facebook (Meta's flagship) had a monopoly on the "personal social networking" market. According to the definition in the brief, a personal social network is an online platform for users to stay in touch with "personal connections in a shared social space."

As stated in the brief, MeWe and Snapchat were the only two direct competitors to Facebook left standing. All other personal social network upstarts had either gone out of business or been acquired by Meta. None of us had upended the emperor.

Surveillance Capitalism has overrun common sense. Fundamental human principles are in peril. In its current state, Web2 is not capable of alleviating its cancers.

I left my day job at MeWe on Independence Day, 2022. New investors and a new executive team there are focused on its success. MeWe is not the solution to this Gordian knot. My mission when I started MeWe was to restore the privacy protections, authenticity, and connectivity that social media was born with. In the past decade, the problems with social media have become much bigger than any single company can solve. The blossoming of issues around mental health, election interference, biased censorship, exploiting kids, boosting, bots, trolls, and much more necessitate a different path forward. A revolutionary framework is required.

2

Uncanny Parallels: Big Ag, Big Energy, Big Tech

To wrap our minds around the Web and its trajectory, it's helpful to observe the patterns of other industries. For our purposes, let's examine food, energy, and the Web. What do these three disparate categories have in common? Each has its own unique issues and peculiarities, but at a high level the parallels are striking, their patterns uncannily similar.

All three are essential in modern society and today are dominated by a handful of monopolistic corporations. Eerie and omnipresent, they have unprecedented power over our lives. Big Ag controls what foods we eat; Big Energy controls what energy we use; Big Tech controls what information we consume.

In the Beginning

One of the challenges in this era of Big Tech is the lightning pace at which innovation and new technologies are presented to us. This speed is historically unprecedented. In the past, a game-changing innovation such as the printing press lasted a few centuries before it was superseded. Social networking was invented in the 1990s and gained mainstream popularity in the 2000s. (Remember MySpace?) Today, over 90% of Americans[1] and 60% of the entire human population[2] actively use social media. In the blink of an eye, it has overtaken our lives and the world.

Concurrent with the tech revolution, there are promising grassroots movements afoot to rectify the problems of Big Ag and Big Energy. There are rapidly amplifying voices for a parallel movement to address the perils of Big Tech. Let's take a closer look at these parallels. What are their foundations? And how did we get here?

Big Ag: The Seismic Shift

Around 10,000 years ago, humans first began to develop agriculture and animal domestication. This was a seismic shift from the nomadic hunter-gatherer lifestyle practiced by our ancestors since the dawn of humanity. From this inflection point emerged the first sedentary tribes, villages, cities, and advanced civilizations. Farming, agriculture, and animal domestication concurrently developed to feed and nourish these growing communities. Yet, over time, advances in technology and new economic models developed that distorted these practices.

For our purposes, let's fast-forward to the mid-20th century. Prior to World War II, there was no widespread industrial farming, also known today as "factory farming." Farms in America were small scale and usually family owned. The stores selling food produced by these farms were typically local mom-and-pops. After World War II, massive, industrialized farms (what became known as "Big Ag") began to arise in tandem with major food retailers ("Big Food").[3] They evolved together to form today's "seed-to-retail-sale" distribution lockstep.

Big Energy: Harnessing Earth's Resources

For thousands of years, ancient people used coal, natural gas, and oil for heating, cooking, and lighting.[4] More than 2,000 years ago, advancing civilizations then began using water wheels to harness the energy of flowing rivers (including the Roman Empire, ancient Egypt, India, and China). Believed to be the first method of mechanical energy to replace the work of people or animals, water wheels were used to irrigate crops, grind grain, and provide drinking water to villages.[5] Next up, windmills came into play over 1,000 years ago. They were used in Europe, the Middle East, India, and China to pump water, grind grain, and assist with food production.[6]

That's enough ancient history. Fast-forward to the Industrial Revolution of the 1800s, when the utilization of fossil fuels like coal and petroleum to power our societies first became widespread. Coal became a critical resource to power steam engines and other machinery. America's first commercial natural gas well was drilled in New York in 1821,[7] and the first commercial oil well was drilled in Pennsylvania in 1859.[8] Since then, fossil fuels have become the primary source of energy across the world, providing us with heat, electricity, and fuel for transportation. While these power sources were originally intended to help us prosper, they've had the unintended consequence of devastating our environment, polluting the air we breathe and the water we drink.

During this period of expansion in the 1800s and early 1900s, there were some who sought alternative paths. In 1884, Charles Fritts, an American inventor, installed the world's first rooftop solar array in New York.[9] This was just two years after Thomas Edison launched the world's first commercial coal plant. A couple decades later, in the early 1900s, visionary scientist and inventor Nikola Tesla made plans for the mass commercial utilization of renewable energy. He envisioned a world with electricity generated via solar, wind, and geothermal technologies.[10]

However, fossil fuels were much more accessible and affordable. They won the day and that early battle, and triggered an unrelenting domino effect. As the 20th century progressed, the energy sector became dominated by a small group of oil and gas conglomerates and utility giants, now known as "Big Energy."

Big Tech: The Birth of the Web and Social Media

The Web and social media didn't emerge until after *Seinfeld* was on the air, yet its precursors were born centuries earlier. Even before Shakespeare. The widespread distribution of targeted information informing and manipulating our philosophical and political thoughts, along with our snake oil purchase decisions, started 600 years ago. The invention of the printing press in 1436[11] was the sea change that launched the Information Revolution. This inflection point of innovation spread throughout the world over a few centuries. Its reign was joined by significantly advancing inventions in just the last 180 years. Examples abound, including photography, telegrams, telephones, film and video, fax machines, and cellphones.

Then came the Web in 1989, eclipsing all that came before. The world went from disjointed to utterly connected in a heartbeat. Like the printing press in its day, this technology is far ahead of any sense or sensibility we have regarding its impact on our lives, minds, and well-being. And hence, a conundrum for regulatory oversight.

While the Web was originally a bridge for communication between university scientists, Web inventor Tim Berners-Lee and his team saw far greater potential. They advocated for the underlying code of the Web to be available to all, free forever.[12] What Berners-Lee didn't realize was that his invention would become the most powerful medium for expanding communication and knowledge in the history of the world. For his unintended stroke of genius, he was knighted by Queen Elizabeth II in 2004.[13]

Inspired by the spirit of Berners-Lee's vision for the Web, early social networks were designed to serve their users as cherished customers. Friends, family, and like-minded folks came together through the magic of this newfangled (and remarkable at the time) communication tech. The wrinkle is that today's lightning pace of innovation partners perfectly with monopolistic capitalist forces. Uh-oh.

Monopolistic Consolidation

So, what has happened? The idealistic foundations of farming, energy, and the Web were overtaken by natural human greed. Over time, this avarice has led to powerful and destructive monopolies. The end results across these sectors bear uncanny similarities.

Big Ag: Factory Farming

Since World War II and the birth of industrialized factory farming, extreme consolidation has occurred across all parts of the food industry. The next time you buy steaks or hamburgers (if you're a meat eater) at the grocery store in the United States, there's more than a good chance they will come from JBS, Tyson, Marfrig, or Cargill. From 1982 to now, these four companies (including their subsidiaries) increased their market control of beef production from about 40% to over 80%.[14] Just four conglomerates also control nearly 70% of all pork and over 50% of all poultry production in America.[15] Similarly, the five largest seed suppliers increased their global market share from about 20% in 1994 to over 60% today.[16]

Giant conglomerates control virtually every aspect of the food production process, from selling feed to farmers to packaging meat and poultry for supermarkets. This situation has led to significantly lower pay for farmers, fewer choices and lower-quality foods for us consumers, and skyrocketing profits for Big Ag monopolies. Bill Bullard, CEO of the Ranchers-Cattlemen Action Legal Fund, stated, "We've never witnessed this level of concentration in the history of our industry. This situation is urgent."[17]

Big Energy: Legal Monopolies

Big Energy includes the major oil and gas conglomerates that fuel most of our freight trains, trucks, ships, airplanes, and gas-powered cars. It also includes the utility giants that supply electricity to our offices, factories, homes, stores, and electric cars.

Let's start with a glimpse at oil and gas. Hello, Model T, good-bye, horse and buggy. In the early 20th century, Standard Oil quickly rose to dominate the booming oil market, eventually owning 90% of all oil production in America.[18] Standard Oil exploited its position by jacking up prices in areas with no competition while lowering prices where competition was active. Standard Oil's executives must have been feeling slick until 1911, when the US Supreme Court ordered it to break up into several smaller companies.

Alas, there's a catch. Many of the smaller companies that were created as a result of the breakup, such as Exxon and Chevron, remained major players and continued to dominate the market. These smaller companies often continued working together to coordinate their actions, which limited competition. In the end, the court-ordered breakup of Standard Oil failed to achieve its goal of promoting competition in the industry.

The industry's consolidation continued unthwarted. In 1998, British Petroleum (BP) acquired fellow oil giant Amoco;[19] in 1999, Exxon and Mobil merged to form ExxonMobil;[20] in 2001, Chevron and Texaco merged to form ChevronTexaco;[21] on and on it goes.

This model has led to us paying higher prices at the pump while having less choice. Additionally, it's hampered innovation since the top companies have far less urgency to compete.

Let's turn to the utility giants. In the early 20th century, the construction of a national power grid required massive investments. To incentivize companies to take on this daunting task, the government granted them the ability to operate in noncompetitive environments. In effect, the utility companies were handed legal monopolies.[22] While this may have made sense in the early 1900s, giving rise to home freezers instead of ice delivery, it doesn't anymore. A century later, we're still saddled with an inbred infrastructure creating barriers to entry.

Big Tech: Enveloping Funding/Monopolistic Consolidation

Big Tech mimicked Big Ag and Big Energy in its transition to insatiable profiteering. In the Web and social media sphere, pioneering upstarts rapidly gave way to a small cluster of powerful conglomerates that quickly gobbled up market share. Following my and others' foray into social networking in the 1990s, new social networks arrived in the early 2000s. These included Yahoo Groups in 2001, Friendster in 2002, and MySpace in 2003.

And then came Facebook in 2004. It's a story with sordid details. Harvard students Divya Narendra and the Winklevoss twins had conceptualized its precursor, "Harvard Connection."[23] The idea was then heisted by their hired gun—fellow student and wunderkind coder Mark Zuckerberg.[24] Shortly thereafter, Facebook was presented to the world as an altruistic platform connecting all of us.

As Facebook spread like wildfire, its true objectives became clear. Rather than serving its users, our favorite geeky villain, Mr. Z, shaped Facebook's business strategy into one big spying eye. It tracked users to aggregate and exploit their personal data through targeting and manipulation, creating the newfangled business model that we now know as Surveillance Capitalism. Today, Meta (which includes Facebook) dominates the social media market with approximately 3.6 billion users across its platforms—nearly half the global population.[25] This footprint is even more staggering considering Meta's ban in huge population centers like China and Russia.

Google rose to dominate its own wedge of the Web along a similar trajectory. Founded in Silicon Valley in 1998, Google achieved stratospheric profits as an early pioneer of the same Surveillance Capitalist business model.[26] The model proliferated across the tech landscape. Today, Google is the unrivaled king of the global search market. Over 92% of all Web searches made across all search providers worldwide are conducted using Google's platform.[27] As of 2024, Google has made 260 acquisitions for a total price tag of over $40 billion,[28] including its purchase of video-sharing giant YouTube in 2006 for $1.65 billion.[29]

Amazon, founded in Seattle in 1994, is another member of the elite monarchy of Big Tech royalty. As of 2023, Amazon represents nearly 40% of the American online retail market.[30] Its closest competitor, Walmart, owns a mere 6%, by comparison.[31] Amazon is also dominant in the cloud infrastructure market—the backbone services of the Web, powering websites across the world. Amazon Web Services (AWS) controls 33% of this entire market.[32] Microsoft Azure is the runner-up with 21% market share, and Google Cloud is in third place with 8%.[33]

How did these nascent Web2 elite of Big Tech afford their rise to power? Through windfalls of funding from venture capitalists and private equity firms. In total, Facebook raised over $24 billion in funding,[34] including $16 billion raised from its initial public offering in 2012.[35] Twitter—prior to Elon Musk's purchase of the company for $44 billion in 2022—had raised approximately $13 billion from investors, including $1.8 billion raised from its IPO in 2013.[36]

These sky-high fundings built a bulwark around Big Tech giants, making them untouchable. Large investors became reluctant to fund smaller upstart competitors. Monopolies were born.

Monopoly Practice: Short-Term Profits, Short-Changed Customers

Big Ag, Big Energy, and Big Tech monopolies share a similar perspective. The consumers of their products—their end users—are not their true customers. These companies all have business models with upside-down incentive structures. In both the short and long term, they harm the very users who they are supposed to serve.

Big Ag: Profits Before People

Big Ag corporations are some of the wealthiest companies in the world. In 2023, Cargill earned revenue of $177 billion,[37] JBS earned over $72 billion,[38] and Tyson Foods earned over $52 billion.[39] As Big Ag consolidated power, it realized it didn't need to market directly to consumers. Instead, its true customers became massive Big Food distributors and retailers like Walmart and Kroger.

Rather than serving end users, Big Ag lines the pockets of its executives and shareholders. The rest of us are left with fewer choices, lower-quality and less-healthy foods, and environmental devastation that we, our children, and future generations will have to pay dearly for.

Big Energy: Profits over Users

Among oil and gas giants, in 2023, ExxonMobil earned revenues over $344 billion;[40] Shell, over $323 billion;[41] Chevron, over $200 billion;[42] and BP, over $213 billion.[43] Among utility giants, in 2023, Duke Energy earned revenues of $29 billion;[44] Pacific Gas & Electric Company (PG&E), over $24 billion;[45] and Exelon, over $21 billion.[46]

As just one example (among many) of Big Energy's misalignment with its customers, let's take a look at PG&E, the largest utility company in California. PG&E was found responsible for the 2018 Camp Fire disaster, the deadliest fire in California's history, due to the company's faulty and unmaintained equipment.[47] Yet PG&E had more than enough funds to perform the maintenance needed. In the five years prior to the Camp Fire, the company distributed over $5 billion in dividends to its shareholders.[48] This is what happens when soulless corporations put profits above all else.

Russell Gold, senior energy reporter at the *Wall Street Journal*, stated, "If PG&E was an individual and not a corporation, I think by now they would be in prison . . . The problem is you can't take a corporation and put it into prison."[49]

Failing to maintain vital equipment is endemic among utility giants, because they have little incentive to maintain it. Due to their monopoly status, their customers frequently have no choice to switch to another service.

Big Tech: Surveillance Capitalism, Users Are Products to Sell

Big Tech companies also put profits first, and there are plenty to be had. In 2023, five titans atop Big Tech's Mount Olympus—Meta,[50] Alphabet[51] (Google's parent company), Apple,[52] Amazon,[53] and Microsoft[54]—raked in a total of over $1.6 trillion in revenue. That's more than the entire GDP of most countries. Data is the new oil, and Big Tech behemoths are the new barons in town.

Big Tech giants powered by the overlay of Web2's Surveillance Capitalism have tossed user privacy and data protection by the wayside. In a curious twist, rather than being customers to serve, we're attractively packaged and on the menu. For example, let's say you're a 25-year-old, love shoes, with a credit card ready to flash. Highest bidder gets to target you. You see, we're the products sold by the network.

The true customers of Big Tech are the advertisers, marketers, massive data brokers, governments, politicians, and any person or entity who seeks to target us to influence our minds, opinions, purchase decisions, and votes.

Monopoly Practice: Turning off the Lights of Competitors

Look at the trails of these monopolies. Big Ag companies use their market power to put small farms and retailers out of business. Big Energy uses its ample resources to dissuade development of renewable energy alternatives. Big Tech uses its formidable heft to snuff out smaller tech upstarts. Behind each is an "out of business" sign in a dark window.

Big Ag's Hollowing Out

Big Ag has put countless small and medium-size farms out of business, leaving a path of hollowed-out local economies across rural America. Unable to compete with Big Ag's mass production machine and cheaper prices, smaller growers continue to wither away. In 1990, small and medium-size farms made up almost half of America's agricultural production. Today, they're less than a quarter.[55]

Big Energy's Carpet Yanking

Big Energy has yanked the carpet out from under countless renewable energy companies, small businesses, and homeowners who want to generate their own energy with solar. Lobbying by utility giants and oil and gas conglomerates has led to significant policy barriers[56] across America that block solar energy's adoption. Big Energy lobbyists have even flooded social media with ads attacking policies that would help the solar industry.[57]

Big Tech's Bulldozed Trail

There's a similar bulldozed trail behind Big Tech. Following the competitive acquisition playbook that we saw in Big Ag and Big Energy, Meta has been on a buying spree for over a decade, including its purchases of Instagram in 2012 for $1 billion[58] and WhatsApp in 2014 for $16 billion.[59] In its pursuit to become social media's Standard Oil baron, Meta even attempted to snap up Snapchat[60] and Twitter/X.[61] However, the CEOs of those companies declined Zuckerberg's advances. In total, Meta has acquired over 100 companies,[62] and likely undermined dozens more.

In 2019, Snap Inc. provided a dossier to the Federal Trade Commission[63] documenting Meta's anticompetitive behavior. Meta's Facebook tentacle cloned Snapchat's features, discouraged content creators from mentioning Snapchat, and blocked Snapchat content from trending on Facebook. MeWe, too, has documented reports from people whose Facebook posts mentioning MeWe were suppressed or flagged as spam. Meta's playbook for wiping out potential threats has been coined "copy, acquire, or kill."[64]

In 2021, the Federal Trade Commission (FTC) filed an amended complaint[65] against Meta for its anticompetitive actions. FTC Bureau of Competition acting director Holly Vedova noted, "Facebook lacked the business acumen and technical talent to survive the transition to mobile. After failing to compete with new innovators, Facebook illegally bought or buried them when their popularity became an existential threat."

A 2023 lawsuit by the US Department of Justice accused Google of illegally stifling competition by paying Apple and other business partners billions of dollars. The purpose was to ensure Google as the default search engine on our Web browsers and phones.[66] A few years back, a 2021 lawsuit alleged that Google gave preferential ranking to YouTube (which Google owns) in its search results and in the Android operating system.[67]

Smaller (but still mighty) Big Tech companies engage in their own version of anticompetitive behavior. In 2022, Twitter/X, under its perpetually fearless new leader Elon Musk, announced its policy to ban links and promotion of Twitter/X alternatives on its platform. Caught in the crosshairs were small upstarts like Mastodon, Tribel, and Truth Social. The policy was later reversed after swift backlash[68] from Twitter/X users. Never one to duck a controversy, Musk then authorized Twitter/X to delay link connections and block link-sharing posts to Substack and media entities like the *New York Times* and Reuters.[69]

Purposeful Misinformation and PR Spin

Big Ag, Big Energy, and Big Tech companies are well aware of the harms they cause. They have all been caught red-handed numerous times promoting false narratives to keep us misinformed about their misdeeds. In their press releases and public statements, Big Ag and Big Energy companies often paint themselves as benefactors of the environment—that's like calling Godzilla a benefactor of Tokyo.

Big Ag's Whitewash

When activists released videos exposing animal abuse in factory farms, Big Ag companies lobbied for new laws[70] to punish those who distribute such images without the facility owner's consent. Holy cow! Big Ag companies were also caught with their fingerprints behind numerous op-eds in major newspapers attacking organic food as an "expensive scam."[71] The US Department of Agriculture even caught Tyson Foods putting false labels on its chicken meat, labeling as "antibiotic free" chickens that were loaded with antibiotics.[72] The deception and manipulation go on and on.

Big Energy's Denial

Exxon (today ExxonMobil) knew about climate change as early as 1977, many years before the issue entered public discourse. Yet into the 2000s,[73] the company vigorously denied its existence. The company reportedly spent over $30 million[74] on purported think tanks that promoted climate denial theories to all of us. The oil giant even hired the same researchers and consultants[75] that Big Tobacco companies used to deceive us about the harmful effects of smoking. Holy smokes!

A peer-reviewed study in 2022[76] showed that major oil corporations, including ExxonMobil, Chevron, Shell, and BP, have all failed to live up to "clean energy" rhetoric their PR and advertising agencies have been spouting for over a decade. These PR promises to become "net zero" have proven to be nothing more than a smokescreen.

Perhaps it's a challenge of becoming too big. Mores fly out the proverbial window as the accountants whittle expenses to favor profits at all costs. The California Public Utilities Commission found that PG&E had falsified its records[77] and committed safety violations for years while misleading the public and authorities. Tragically, in the aftermath of California's Camp Fire in 2018, the company pled guilty to 84 counts of involuntary manslaughter—caused by the behemoth's negligence.

Big Tech's Spin Cycle

In a curiously similar parallel, Big Tech companies have massive teams of PR pros and communications experts that they deploy to skillfully shape and distort our thoughts and opinions. These companies publicly pat themselves on the backs as heroes of privacy while surreptitiously slurping up our data and selling it to the highest bidders.

Meta hired a PR firm to downplay and trivialize the significant scope and impact of Russian disinformation campaigns that infiltrated many of our newsfeeds during US elections.[78] Google has repeatedly told us that it doesn't collect our data from Gmail.[79] It even promoted and fed us a brilliantly deceptive motto for years: "Don't Be Evil." This is just laughable. Google has spent years watching and targeting us based on our purchase histories, locations, searches, links we click, topics in our emails, and more. In a de facto confession, Google deep-sixed its "Don't Be Evil" masquerade in 2018.

The tech giants' PR spins are designed to continuously confuse us. They make us believe in an alternative reality where they are good citizens of the world who have our backs. The fact that media companies now rely on tech giants like Meta, Google, and Twitter/X for the distribution of their stories further complicates their symbiotic relationship.

Many news publishers receive most of their digital traffic from these tech giants.[80] It's unrealistic for a local publisher, like the *Chicago Tribune* or *Houston Chronicle*, to demand payment from Google or Meta in exchange for visibility in search results or users' newsfeeds.

Today, over 70% of Americans[81] say they get news from social media—you probably get some of your news there too. These media giants can pull the levers to decide which stories you see and which you don't. This fact contributes to the manipulation of our thoughts and opinions from behind their flashy curated (based on our data) newsfeed curtains.

Here's what tech journalist Jacob Silverman said[82] about Meta's entanglement with the media: "What Facebook has become is the press's assignment editor, its distribution network, its great antagonist, devourer of its ad revenue, and, through corporate secrecy, a massive block to journalism's core mission of democratic accountability."

Armies of Lobbyists

You might be wondering why aren't there stronger regulations to protect us from the rampant harms caused by these industries? Big Ag, Big Energy, and Big Tech companies all maintain their dominance by spending fortunes lobbying politicians and governments. Their goal? To neuter strong regulation and entrench their power further.

These companies have revolving doors between their executive boardrooms and the regulatory committees meant to rein them in. As just one example among many, take a look at Joshua Wright. Google funded his academic research at George Mason University before he became a commissioner at the FTC. Soon after his government tenure, he hopped right back onto Google's payroll as a key advisor and attorney.[83] This kind of double agent side-swapping belongs in a James Bond movie. Alas, welcome to the reality of Big Tech/Government incest.

Big Ag Spends Billions

Over the last two decades, Big Ag has spent over $2.5 billion on lobbying[84] in the United States. In 2023 alone, the industry spent over $177 million.[85] Additionally, in the last two decades, it's estimated that Big Ag companies donated over $750 million to political candidates.[86]

Tyson Foods alone averages a donation of over $300,000 to political campaigns each election cycle. Big Ag companies like JBS and Cargill spend millions on lobbyists and political contributions as well.

Big Energy Bribes Public Officials

Since 2008, the oil and gas industry has also spent over $2.5 billion[87] on lobbying in the United States. In 2023 alone, the industry spent over $132 million.[88]

As if spending millions on lobbying and political contributions weren't enough, some Big Energy companies have even been caught bribing public officials outright. In 2020, Exelon subsidiary ComEd agreed to pay $200 million for its bribery of public officials in Illinois.[89]

Big Tech Spends Hundreds of Millions

Big Tech companies employ hundreds of lobbyists, overtaking the traditional top spenders, such as the oil and tobacco industries. In 2023 alone, Meta and Amazon each spent over $19 million on lobbying;[90] Google spent over $14 million;[91] and Microsoft and Apple each spent roughly $10 million.[92] Between the 2020 and 2022 election cycles, these five Big Tech companies set new records by collectively spending over $150 million on lobbying and campaign contributions.[93] Of congressional members with jurisdiction over privacy and antitrust matters, 94% have received contributions from a Big Tech political action committee or lobbyist.[94]

The Damaging Impacts of Big Ag, Big Energy, and Big Tech

The closed businesses and barren competitive landscape aren't the only aftereffects of consolidation. The handful of monopolies that now dominate farming, energy, and social media are causing real-world harm. Looking at the impacts of each, their parallels again become clear.

Big Ag's Unhealthy Side Effects

To be fair, there are bona fide benefits from industrial factory farming, including larger harvests per acre with less labor inputs. The overpopulated world can be fed. Farming in this manner can be more efficient and cost-effective and can offer cheaper prices. However, the side effects and unintended consequences of industrialized farming are also bountiful.

Today, a small group of powerful Big Ag corporations are decimating small farms and gutting rural America. These corporations destroy natural habitats, deplete natural resources, diminish biodiversity, degrade farmland, and accelerate climate change. Our food systems—largely dominated by Big Ag—account for more than one-third of all global greenhouse emissions.[95]

Big Energy: Inconsistent with Human Survival

Like Big Ag, Big Energy has its benefits. These companies provide a reliable source of energy that is abundant, affordable, and versatile. It is used to power everything from our cars and planes to factories and generators of electricity. Fossil fuels have also been an important source of revenue and economic development for many countries. Those countries make unholy deals with Big Energy to reap needed funds while ignoring the ensuing resource pillage. As our planet broils, the consequences are clear.

In a speech in 2023,[96] Secretary-General of the United Nations Antonio Guterres said: "Today, fossil fuel producers and their enablers are still racing to expand production, knowing full well that this business model is inconsistent with human survival. Now, this insanity belongs in science-fiction, yet we know the ecosystem meltdown is cold, hard scientific fact."

Big Tech's Overarching Dark Side

Like the other giants, Big Tech companies such as Meta, Google, and Amazon have many benefits. They offer us a wide range of products and services that make it easy and convenient for us to access information, communicate with our friends and family, shop, and conduct business. Web and social media companies like Meta and Twitter/X have allowed us to connect in ways that were never before possible. Google has made information more accessible than ever. The Big Tech companies are often at the forefront of innovation, and they are major drivers of economic growth. Yet Big Tech has a dark side that would make Darth Vader blush.

Meta has allowed violent extremist groups,[97] from Mexican drug cartels to ISIS to Hezbollah, to proliferate and recruit on Facebook. The social network has even been used to incite genocides in Myanmar.[98] The *Wall Street Journal* revealed an internal Meta report showing that 64% of users who joined an extremist group on Facebook did so because Facebook's algorithms recommended it to them.[99] That's crazy.

But there's more. As reported by *The Guardian* in 2023, Meta (via Facebook) was responsible for nearly 95% of the 29 million reports of child sexual abuse material online in the prior year.[100] No words . . .

Meanwhile, Twitter/X allows known terrorists to purchase verification checks on their profiles to boost their content across the platform, in potential violation of US sanctions, according to a 2024 report by the *New York Times*.[101] The company is known to "trend" (boost to the top of user newsfeeds) inciting content—for example, trending "Hang Mike Pence" to millions of its users for several hours on January 6, 2021.[102]

Big Tech companies have assailed our privacy, squashed countless tech upstarts, decimated local news companies, engaged in biased censorship, disrupted our democracies, amplified hate and misinformation, and fomented a mental health crisis like the world has never known.

Fines Are Ineffective

Big Ag, Big Energy, and Big Tech companies get more fines than NBA players. Yet with pockets deeper than the Mariana Trench and vast armies of lobbyists and lawyers on the payroll, these fines are nothing more than parking tickets.

Big Ag's Wrist Slap for Manipulating Prices

Big Ag has collectively been ordered to pay hundreds of millions of dollars in fines in just the last several years. For example, in 2021, Tyson Foods agreed to a $99 million settlement as part of a class-action lawsuit over chicken price-fixing.[103] Accusations against Tyson include participating in a nationwide "Chicken Cartel" with America's top chicken processors. They deliberately reduced supplies by culling flocks of breeder hens, and then jacked up chicken prices in the market. This was far from the first time the food giant has had egg on its face.

Similarly, Dairy Farmers of America (DFA) was forced to pay $50 million in 2016 due to allegations it conspired to limit competition and manipulate prices.[104] Considering net sales that year were $13.5 billion,[105] it was more than willing to pay this nominal fee—you won't see DFA members crying over spilled milk.

Big Energy Fined Billions

Things look pretty much the same with Big Energy. The sector has accrued tens of billions of dollars in fines in just the last decade or so. Case in point: In 2015, BP agreed to a $20 billion settlement over its 2010 oil spill that ravaged the Gulf of Mexico.[106] This was hardly BP's first penalty for causing environmental disasters—the company knows the drill. Yet as it soaks up cash by the barrel, such fines are a drop in the well. In 2023, BP's net profits were over $15 billion.[107]

PG&E has been fined nearly $3 billion for its laundry list of mishaps, including a whopping $2 billion fine for its role in Northern California's catastrophic 2017 and 2018 wildfires.[108] That amount might sound shocking, but PG&E's legal monopoly over much of the state gives it money to burn. In 2023, the company saw net profits of over $2 billion.[109]

Big Tech Avoids Trillions

It's certainly no surprise that Big Tech companies are fined frequently too. But these giants simply ignore the regulations and write off fines from the FTC and governments worldwide as costs of doing business. Meta was fined $5 billion by the FTC in 2019 for mishandling its users' personal information in the wake of the Cambridge Analytica scandal.[110] Yet the *Washington Post* estimated that since all American users may have had their data scraped, the maximum possible fine the FTC could have imposed on Meta was approximately $7.1 trillion.[111] That's a tad more than $5 billion. For reference, according to the Federal Reserve, there is just over $5.8 trillion US money in circulation, total.[112] It's telling that Meta (which had earned $15 billion in revenue in the previous quarter alone) saw its stock price surge immediately after the FTC's fine was announced, a surge that more than covered the fine.[113]

Big Ag, Big Energy, and Big Tech are ablaze with examples of capitalism run amok. The Godzillas are on a shackle-free rampage. Can we save Tokyo? Efforts are afoot. Turn to Chapter 3 and check out the restorative movements going toe to toe with the reptiles.

3

Promising Movements Afoot

AMID THE PROFITEERING binge, promising grassroots movements are gaining traction to fix the reptilian problems of Big Ag and Big Energy. These movements provide a glimpse of what's possible in a parallel movement to transform Big Tech.

Big Ag: Regenerative and Organic Farming

"Regenerative farming" is a term you might have heard before. This now-blossoming movement to fix the farming sector began with a few pioneer farmers in the early 1980s.

Regenerative farming is a holistic approach to farming that centers on the interdependence of farming systems and their ecologies. For example, plants such as grass and hay are grown locally on the farm to feed the livestock, then the manure produced by the livestock is used to fertilize the plants. Regenerative farms use techniques such as crop rotation, effective bed planning, and maximizing the benefits of natural processes, like microbial relationships with plants. This approach preserves the health of the soil and, unlike industrial Big Ag farms, most regenerative farms are either carbon-neutral or carbon-negative.

Gabe Brown, owner of the regenerative Brown Ranch in North Dakota, says, "Regenerative farming is getting the farmer or rancher closer to the consumer, and those consumers are willing to pay for that. So, it's putting more money in the pockets of farmers and ranchers and putting a higher quality product onto our tables."[1]

In tandem with the regenerative movement, the organic farming movement is blooming as well. Many regenerative farmers also use organic farming practices.

Worldwide, currently over 100 million acres of land are dedicated to regenerative agriculture. Based on current trends, climate database Project Drawdown estimates that number will grow to 1 billion acres by 2050. This transition would remove 23 gigatons of carbon dioxide from the Earth's atmosphere.[2]

Big Energy: Renewables and Electric Cars

In parallel, there is a growing movement to fix the problems of Big Energy via renewable energy. People around the world are installing rooftop solar systems on their homes to reduce their energy bills. In doing so, they gain independence from the utility giants and support the environment by slashing their carbon output.

The first commercially available solar array was introduced in 1956,[3] but the cost of $300 per watt made it inaccessible to the public. (Adjusting for inflation, that's over $3,000 per watt in 2024 dollars.) Today, a solar array can cost as little as $0.50 per watt.[4] This change showcases astonishing possibilities. Advances in battery storage and supportive government policies now make rooftop solar a viable option for millions of households.

One of the most important government policies aiding the rise of solar is called "net metering," which allows residential and commercial customers who generate solar electricity to sell any excess back to the grid.

According to a 2023 report by the International Energy Agency, solar, wind, and other renewables are expected to account for 35% of the world's electricity (and overtake coal) by 2025.[5] The clean energy promised land is within our reach.

Instead of Big Energy conglomerates leading the way, the renewable energy movement was led by outsiders, including solar companies like First Solar, SunPower, and SolarCity (which was later acquired by Tesla).

Similarly, the electric car industry was led by entrepreneurs rather than established giants. General Motors took a stab with its launch of its electric car, EV1, in 1996. Despite its passionate owners, GM's efforts were questionable at best, culminating with the company recalling all EV1s and crushing them.[6]

It took an outsider like Elon Musk and Tesla to fire up this opportunity. Electric vehicles sales have grown from next to nothing a decade ago, to 18% of global car sales in 2023.[7] Major car companies are now putting electric vehicles at the center of their business plans, and many have signed pledges to eliminate sales of gas and diesel cars globally by 2040.[8]

Regenerative farming, renewable energy, and electric cars all began as niche movements embraced by small numbers of true believers. Were you one of them? It's fun to be an early adopter . . . and challenging! In recent years, these movements have entered the mainstream. When tasked with solving big problems, new startups with entrepreneurial spirits step up to the plate in unprecedented ways.

Where Does This Leave Big Tech?

The good news starts with how fast tech evolves today. Moore's[9] and Koomey's laws[10] and other guideposts document this rapidity. Reflecting their technology chops, the regenerative and renewable movements disrupting Big Ag and Big Energy are also now iterating at unprecedented paces.

The time is ripe for a comparable movement to transform Big Tech and supersede Surveillance Capitalism. Swooping in to save the day, Web3 is the anointed superhero.

Has the Big Tech villain been vanquished?

4

Web3 Is Here: What the Heck Is It?

You've likely heard the term "Web3." But what the heck is it? It's mystifying to most of us. Its missionaries dub it the next generation of the Web. They evangelize a digital utopia where all the problems of social media and the Web are resolved. How? Via blockchain and token-based technologies that herald a decentralized and open internet. They argue it will foster a democratic and equitable future. We as individuals will not be manipulated or targeted, and we will have total control of our data and online communities. Sounds lovely, eh?

But wait—what do we mean by centralized versus decentralized? Centralized platforms (like the Web2 social sites we use) each possess and maintain our content and data that we post there. They have complete access. Think of a tree. There can be lots of different branches, but they're all rooted in one central trunk. In contrast, on decentralized platforms, our data is spread across numerous private, anonymized nodes.

Bertrand Perez, chief operating officer of the Web3 Foundation, shared the vision this way: "The idea of Web3 is to decentralize the data so no single organization will have full control. That will provide you more freedom because you don't need to trust a single organization."[1]

While the Web2 era transformed us into the products sold to advertisers, Web3 proclaims it flips this paradigm on its head by converting us regular users from "products" into "owners." In this vision, you have ownership of the content and data you create and share online.

Right now, power over the Web is concentrated in the hands of a few giant corporations. It's troubling that a small, elite group of Big Tech corporations, cemented in power by anticompetitive actions, have unchecked reign over the Web.

Web3 proponents say it will redistribute the power, taking it away from these megacorporations and into the hands of regular users like you. How did we arrive at the white sandy beaches of this purported promised land? What are the tangible differences among Webs 1, 2, and 3? Let's revisit the past with a sharp lens to grasp how we arrived at Web3.

Web1

The year was 1989. Tim Berners-Lee, a brilliant computer scientist who I'm honored to call a friend and advisor, was working in a lab at CERN, the European Organization for Nuclear Research, in Geneva, Switzerland. It was there that he developed the first protocols for what would become the World Wide Web.[2] His vision was to create open, decentralized protocols that enabled sharing information between computers from any location on the planet. This earth-shattering invention heralded Web1, the first era of the Web, lasting from its birth until the early to mid-2000s.

In early Web1, commonly dubbed the "read-only Web," most websites were static and owned by companies. There was little interaction or contribution from users; it was much harder to produce content of your own. Web users were "consumers" who visited websites to consume information. If you or I visited the same site, we would see the same content, presented identically.

Piggybacking off Berners-Lee's invention, the latter half of Web1 saw the invention of the world's first social networks in the 1990s.[3] Budding social media made it possible to stay connected with friends, family, coworkers, and like-minded people around the world. It was altruistic and idealistic.

This would all change in Web2.

Web2

In the early to mid-2000s, we entered Web2—the iteration we currently live in. The biggest evolution between Web1 and Web2 was interactivity. In this era, the Web transformed from mainly "read-only" into a “read-write” interface. While social media existed in Web1, it didn’t hit mainstream audiences until Web2. Companies were no longer the primary content providers; users could now generate and share content. New, highly interactive sites came on the scene, like Facebook, YouTube, and Twitter/X. Online platforms focused on facilitating these user-to-user interactions.

With increasing numbers of people accessing the Web, a small group of dominant companies quickly rose to claim a disproportionate share of the Web’s traffic and value. These centralized platforms started collecting data about you, under the guise of serving you better content. In this version of the Web, not only do you get information from websites, they also get information from you. Valuable information. These companies soon realized they could package the massive troves of data they had on you and sell access to any advertiser that wished to target you.

Web2’s business model is known as Surveillance Capitalism, and we are living under its unblinking eye.

Welcome to Web3

So here we are, on the frontiers of Web3. Web3’s foundations are in blockchain, tokens, and decentralization. It all starts with blockchain. You’ll learn more about tokens and decentralization as we progress.

But what exactly is blockchain? Let me give you the quick-and-dirty *Cliff Notes*.

A blockchain contains a peer-to-peer digital database—like a ledger on the Web—that is distributed across a network of computers. These computers are called “nodes.” Think of these nodes as the vertebrae and the network as the spine. This database is secured using cryptography, a method of concealing information using codes so that only people who are intended to can read it. Each transaction in this database is linked to the previous record. This creates a chain of records (which are called “blocks”). Hence, the blockchain.

On Halloween 2008, an anonymous person, or perhaps a group of people, acting under the pseudonym Satoshi Nakamoto, published a white paper online titled “Bitcoin: A Peer-to-Peer Electronic Cash System.”[4] This document introduced and delineated the concept of Bitcoin, the world’s first blockchain and cryptocurrency. The idea was to establish a secure, decentralized, and transparent system for storing and transferring digital currency—Bitcoin in this case—without requiring a bank or centralized authority.

The purpose of our inquiry is not to assess the viability and value of cryptocurrencies. That's not my expertise. Our Web3 undressing will focus on this new Web's underpinnings.

While Bitcoin forged a path into Web3, it was built specifically for the narrow purpose of exchanging Bitcoin cryptocurrency. It could not be used as the foundation for Web3 apps and services. Consequently, numerous blockchains have popped up in Bitcoin's wake to fill this void. These blockchain-based technologies are used for various applications beyond cryptocurrency, such as supply chain management, healthcare, decentralized finance (DeFi), real estate, digital art, games, and more.[5]

Gavin Wood, cofounder of Ethereum, the second largest blockchain and cryptocurrency behind Bitcoin, is said to have coined the term "Web3" in 2014.[6] Ethereum is now the most popular blockchain platform on which Web3 apps and services are built.

The Promised Benefits of Web3 for Social Media

For several years, social media has been envisioned as a key beneficiary of Web3's milk and honey. Imagine a decentralized and well-governed social network where you own and control your data, content, identity, and destiny. In this ecosystem, issues of targeting, bots/trolls, content ownership, portability, and moderation are resolved. Regular users like you oversee governance, moderation/censorship decisions, and oversight, incentivized by a foolproof crypto-based rewards system.

The list of purported benefits is long and dazzling, paving the way, it seems, out of the surveillance model of Web2. Let's investigate some of the promised upsides of Web3's social media platforms.

Single Sign-on

Today, many of us use our Facebook, Google, or Apple logins to easily sign into new apps and services across the Web. However, when you use these sign-in services, sometimes referred to as "Big Login," you are unwittingly sharing your personal info across the Web. This makes it easier for Big Tech giants to target you.

Web3 promises an alternative solution, "Single sign-on" (sorry, it's not a newfangled dating app). This allows users to log in to various Web3 applications using a single set of login credentials. Here is where their crypto wallets come in. (Don't worry, we'll dive deeper into wallets later.) For now, it's helpful to know that these wallets serve as a user's central authentication point. They hold the user's private keys, which are like a digital signature used to log in to applications and services across the Web3 world.[7] All without having to give up data to Big Tech.

Data Ownership and Portability

Web3 claims to give users complete ownership and control of their content and data. Users have full autonomy over their online presence, with the ability to own and transfer their content and data anywhere around the Web with ease. Ownership of virtual items—from social media posts and avatars to songs and digital artwork—are placed in the hands of users through nonfungible tokens (NFTs).

The next puzzle piece is an easy way to move your data and content between platforms. In 2024, if you want to migrate from Facebook to Twitter/X, TikTok, or elsewhere, you can't take your content or your relationships with you. Web3 promises the ability to easily move from platform to platform and service to service, taking your content and contacts/followers with you.

A quick aside—true fact: Big Tech giants have feigned plans for data portability in the past. In 2018, Facebook, Google, Microsoft, and Twitter made waves by announcing the Data Transfer Project. Indeed. Their claim was that they would allow their users to seamlessly transfer photos, contacts, mail, and other data within their closed clique. This was a shameless attempt to pacify the antitrust eagles circling overhead. Of course, those companies weren't direct competitors, and their plan would only build a moat against smaller upstarts. Ultimately, the only thing accomplished by their big announcement was a brief cycle of positive PR. Their pretend portability plan never materialized.[8]

Reed McGinley-Stempel, CEO of software development firm Stytch, says, "Data portability was an afterthought in Web2, but it's a first-class citizen in Web3. There's nearly no end-user friction involved in porting your data from one service to another in Web3."[9]

Secure Data Storage

So, where exactly does your data reside in Web3? Whose server is the guardian of your family photos? According to Web3 development firm Moralis, files on Web3 "are stored on a network of computers rather than on a single server."[10]

In Web2's current storage landscape, nearly all apps and sites use just three companies to store our wealth of data: Amazon Web Services, Microsoft Azure, and Google Cloud. There our data is centrally controlled and accessible to the peering algorithms of the social sites we use.

Rather than having three megacompanies store your data in one or more of their massive data centers, Web3 entities prefer a different type of storage and fragmentation. Shards of your data are stored across numerous computers whose owners are monetarily incentivized with tokens to rent out free disk space on their drives to host it. With your data stored across numerous computers, or "nodes," around the world, redundancies are provided so that there's no single point of failure. If one server goes down, there will still be plenty of others with the same parcel. Web3 companies typically partner with third-party decentralized storage companies like Filecoin and Arweave.

How is your data kept secure when it's hosted on the computers of strangers around the world? That's where cryptography steps in. In decentralized storage, your data gets parsed into little pieces and encrypted. The data then is dispersed into secure, private vaults in various nodes across the globe. Each of these vaults has a key only its owners have access to. Even if hackers could make it into these vaults, they have access only to tiny pieces of your data, not your whole treasure trove, because your data is safeguarded across so many of these private vaults. For this reason, decentralized storage is difficult to hack.

Crypto Rewards and Democratized Moderation

One of the key features of Web3 social media platforms is the ability for you to earn cryptocurrency rewards for your contributions.[11] Web3 networks typically have their own native cryptocurrency (while some use more established crypto such as Ethereum). You can earn the crypto by creating content, commenting, voting on other users' posts, or by helping to moderate the platform. You can then use these tokens to pay for enhancements and bonus features on the site, and in some cases cash them in for fiat—government-issued money: dollars, pounds, pesos, and so on.

Imagine this scenario: You log in to Instagram. You're scrolling through, and up pops a post from an old high school acquaintance. They take a stance on the latest hot-button political issue, and like being hooked on a slot machine, you load "more" to see the comments. There it is before you, humanity's worst traits. The whole comments thread devolves into name calling and an ALL-CAPS shouting match. Like watching a car wreck, it's horrible, but you have a hard time looking away.

Web3 promises to solve this. Its token rewards system purportedly incentivizes you to produce high-quality content and engage positively with others. In doing so, you receive crypto tokens from your peers—a reward system for humane online behavior. This is Web3's solution to online moderation.

Eliminating Bots and Trolls

Web3 proponents proclaim that their new version of the Web helps eliminate bots and trolls. Many Web3 sites require a cryptocurrency transaction to interact with them, which can come with associated fees. It's argued that these fees would prevent large-scale automated bot activity, since the purveyors of bot armies wouldn't be willing to pay. Ned Scott, cofounder of Web3 platform Steemit, has described how on Steemit and similar Web3 platforms with token reward systems, "there's an opportunity cost for trolling."[12]

Data Privacy, Anonymity, Freedom from Targeting

With decentralized storage and cryptography, Web3 boasts that its users enjoy data privacy and anonymity par excellence. Platforms supposedly cannot exploit your data, target you themselves, or share/sell your data to advertisers and marketers who seek to target you.

So far, Web3 appears marvelously inviting. But is it really the paradigm-shifting Web revolution it's cracked up to be? Let's dig deeper.

5

A Healthy Dose of Web3 Reality

The benefits of Web3 we discussed in Chapter 4 sound effervescent. But before we jump headfirst into those crystal, bubbly blue waters, let's put magnifying goggles on. Single sign-on, data ownership/portability, and secure data storage are excellent upsides. Yet each of these can be readily achieved without Web3, as we'll discover in upcoming chapters. So, what about all of Web3's promises?

Tone Deaf: Crypto Rewards for Posts and Democratized Moderation

Let's step back from the promised land of Web3 to what mainstream social media users are doing on Web2 platforms. Users of full-featured social networks use these sites for essential communication—interacting with friends and family, engaging in groups, following pages, chatting, and posting stories, disappearing content, and more. This is not transactional. Underneath the jazzy, ever-advancing tech features is simply human-to-human interaction.

Web3's idea of monetizing our communications with a cryptocurrency rewards system for upvoting, downvoting, and moderating posts can work on sites like Reddit or YouTube. Those aren't personal connection sites. But Web3 deploys this complex interactive financial-incentive model for personal social networking as well. Therein lies the rub.

While this idea might sound good on paper (who wouldn't want free money for posting?), there is a disconnect between the inherent problems of social media and Web3's solution.

Incentivized moderation in Web3 is a bit of a pickle too. Crypto rewards can be effective on a small scale, but pitfalls abound. For example, if a Web3 social media platform is an echo chamber of hate, violence, prejudice, and bullying, the moderation model collapses as everybody is in it for the same thing.

The decentralized nature of Web3 creates a double-edged sword. While it can allow for more freedom of expression compared to the Web2 Big Tech giants, it can also make it impossible to get rid of seriously toxic content. Eugen Rochko, founder of the decentralized social network Mastodon, told *The Verge* when asked about hateful and violent content on his platform: "It's not actually possible to do anything platform-wide because it's decentralized. I don't have the control."[1]

The permanence of the blockchain presents a real conundrum for content moderation. If someone encodes a text string with a URL for a website that has child pornography or any other kind of illegal activity, it cannot be removed or altered. The extraction of such data necessitates technical knowledge and effort. Even if the linked website is shut down, the text string persists forever.[2] Additionally, it's possible to engage in harassment on the blockchain via messages that cannot be edited or deleted.

Proponents of Web3 grip tightly to the notion of public governance—the idea that a community of users will self-moderate effectively at massive scale. But this idea seems naïve at best. There is no clear answer yet to how violence, bullying, harassment, and other toxic content can be dealt with effectively at scale on Web3.

Vili Lehdonvirta's 2022 book, *Cloud Empires*, did a brilliant job illustrating the inherent flaws of this kind of decentralized governance decades ago. It highlighted the rise and fall of Usenet, a decentralized online community in the 1990s.

> When the community was small and the membership stable, transactions functioned reasonably well without a centralizing authority . . . However, when the number of users increased, chaos gradually established itself within that community, generating distrust and ultimately the disintegration of that space.[3]

Déjà vu?

This approach reminds me of the problems inherent to direct democracy within politics, which often enables majority groups to discriminate against minorities. The United States and other democratic nations support representative democracy to solve this issue. In social media, as in our political affiliations, users choose the platforms that best represent their values. Direct democracy doesn't work for nations, and it doesn't work for moderating complex social networks at scale either.

Additionally, if a social app wants to achieve mainstream success, it needs to be available on the Apple and Google Play app stores. These app stores are the gatekeepers to global distribution and require more than a modicum of oversight.

Fat Chance: Eliminating Bots and Trolls

Winning the battle against bots is vital to fix social media on any version of the Web. How do Web3 platforms really fare in this fight? Since no Web3 platform has achieved scale on par with the current Web2 social media giants, no one can say for sure. Yet even without knowing their at-scale bot-fighting capabilities, Web3 platforms already have several troubling issues to contend with.

Bots are already running wild on Web3. In the gaming sphere, bots reportedly make up an average of 40% of the user base.[4] The problem is endemic across Web3. For example, a report by Levan Kvirkvelia, cofounder of Jigger, an anti-bot protection software, found that bots make up 87% of the user base of Ariva Digital, a blockchain solution for tourism; 84% of AnRKey, a social crypto platform; and 65% of Voxels, a virtual world on the Ethereum blockchain.[5]

Kvirkvelia explained that "all services with a profit are flooded with bots." His reporting shows how it's surprisingly easy for one person to create multiple crypto wallets on a network. Web3 platforms are designed to reward users with monetary tokens for their activity are a perfect environment for bots to game the system.

Crypto newsletter *Milk Road* reports:

> Web3 activity is significantly propped up by hordes of bots . . . Bots have been plaguing crypto projects for years, spamming the comments of leading figures in the space and spreading scams across

> the community. In the NFT [nonfungible token] market, bots are further used to automatically mint new NFTs, which are then flipped for a profit. Even in some major NFT auctions, so-called NFT sniper bots monitor auctions and submit last-moment bids, buying up limited-edition digital assets at a low price.[6]

Web3 fairs no better in deterring trolls and other bad actors either. Web3 platforms are frequently preyed on by hackers, scammers, and fraudsters. These malicious tricksters swindle billions of dollars from Web3 networks and users each year, including nearly $6 billion of cryptocurrency lost between 2022 and 2023, according to Chainalysis.[7] Members of the Web3 community are working on remedying this ongoing problem, but it's a significant hurdle that's far from solved.

There's another fly in the ice cream: boosted posts. On many Web3 social media platforms, users can stake their tokens to boost their posts or bump up their comments to the top of comment threads. For instance, before Voice pivoted from a social network into an NFT marketplace, Block.one chief technology officer Dan Larimer trumpeted the platform's token-based system in which "users could stake Voice tokens to move to the top of a chain of comments."[8]

This is exactly what happened on Twitter/X. Musk allowed any user to purchase a verification badge; with that, their posts received priority positioning in newsfeeds and comment threads.

This kind of tokenized boosting system is a gold mine for exploitation. Along with individual bad actors, bot and troll armies are often funded by foreign state entities seeking to disrupt civil discourse.[9] (See Chapter 16.) In other words, these entities receive their own wallets and the ability to use tokens to feature their posts in front of a broad audience. Thus, they have even more power to amplify their disruption.

Paid boosting is one of the key ways bots and trolls spread their poison on Web2 giants like Facebook and Twitter/X. Anyone who wants to can buy their way into your feed. Today Web3 does not solve this problem; it amplifies it.

For example, Web3 and decentralized social networks like Minds, Steemit, and Mastodon all have "trending" posts and hashtags, served to you from users and entities you did not choose to follow.[10] Bots, trolls, and other bad actors can exploit these features on Web3—just like they do on Web2.

Debunked: Data Privacy, Anonymity, Freedom from Targeting

How does Web3 handle privacy, anonymity, and targeting? There's a great deal of mythology here that needs busting.

When data privacy is the objective, it's a no-brainer to avoid Web2 Surveillance Capitalists. Along the same lines, it's common sense to think twice before placing your personal data on a public blockchain (remember, they're all public) that's immutable and accessible to any person and entity on the planet.

Are there Web3 social networks that collect and share your data with advertisers, marketers, and other third parties? In short: yes. For instance, if you read the fine print of the privacy policy of Web3 YouTube alternative Odysee,[11] you'll find that they track your information when you do just about anything on their site. Their reasoning? To improve your experience—the same excuse Web2 companies give. As part of their "third-party disclosure" policy, Odysee writes that "non-personally identifiable visitor information may be provided to other parties for marketing, advertising, or other uses." Many other Web3 and decentralized social sites have similar data collection policies. So much for data privacy on Web3!

What's the skinny on anonymity? One of the primary functions of the blockchain is that it makes transactions transparent and trackable to guarantee trust for all parties involved. This visibility presents a significant conundrum for anonymity on Web3. Regulatory compliance attorney Jamilia Grier warned, "In a blockchain transaction, your data and information will be in the transaction. Transactions can be tracked down all the way to their sources."[12]

If you can be connected to your posts and interactions in Web3, then there goes the whole promise of anonymity. As reported by *Business Insider*, "Bitcoin anonymity is just a big myth";[13] the same goes for other blockchains Web3 is built on.

Its supporters proclaim that even though data stored on a blockchain is both public and permanent, it is ostensibly protected through cryptography and pseudonymity. This means that transactions on a blockchain are linked to addresses or public keys rather than to real-world identities. However, experts warn that it's possible to trace someone's activities on the blockchain through pattern analysis or by linking their blockchain address to their real-world identity.

Blockchain expert Feng Hou, Maryville University's digital transformation chief, said that when it comes to crypto, "Nobody can claim 100% anonymity . . . through forensics analysis, we can always get to the bottom of it."[14] According to Ben Weiss, CEO of crypto ATM operator CoinFlip, "Bitcoin transactions are more traceable than cash."[15]

Research at MIT revealed that it often takes only four data points, specifically the dates and times of purchases, to determine someone's real identity.[16] In the MIT study, from a database of credit card transactions over 30 days by over a million people, the researchers were able to successfully identify 90% of the individuals. Think about that. Any personal transaction you make or interaction you have on the blockchain becomes permanently enshrined public data for all the world to see.

It all leads to the Holy Grail: targeting. If your social media data is on the blockchain, where it can be traced and tracked by third parties, what's to stop advertisers, marketers, social media companies, data brokers, governments, and any other entity from accessing it and targeting you? Not a whole lot.

As we've noted, many Web3 social networks take an ad approach right out of the Web2 handbook, giving their users and advertisers the option to pay to boost posts. What's the upshot? Users on these Web3 platforms are getting targeted by unsolicited ads in their newsfeeds that they didn't choose to see. For all its attempts, Web3 is falling flat in creating an advertising model that protects and respects users.

Tectonic Trouble

While there are countless issues to resolve for a happily united marriage of Web3 with personal social networks, there are also deep ground-level problems. Disappointingly, it appears that, currently, both Web3's engineering premise and economic foundations are flawed for social media applications. Here are a few of those flaws.

Nuts and Bolts: Serious Infrastructure Flaws

Web3 relies on your computer and mine for decentralized data storage. This data has to be called into action time and again. Can the decentralized data, blockchain-based Web3 provide seamless, instantaneous delight on par with the current options? Web2 glitters with a broad array of social features—photos, videos, stories, messaging, newsfeeds, music, disappearing content, pages, groups, music, livestream, and more.

Web2 social networks function at lightning speed and process majestic numbers of requests from hundreds of millions of users, simultaneously, in nanoseconds. There's simply no evidence yet that blockchain-based Web3 can pull this off. Its infrastructure does not function nearly as well—or quickly enough—for vastly interactive social networking.

Notable detractors dismiss Web3 as a Silicon Valley fad flawed from the get-go. Some, like Elon Musk (Twitter/X's owner since 2022), have toyed with the idea of creating their own versions of a decentralized network to rival the Web2 giants. These plans were abandoned when it was realized that such endeavors are bound to fail because of Web3's inadequate and molasses-slow infrastructure.

During the 2022 Web Summit tech conference, Web inventor Tim Berners-Lee said, "Ignore the Web3 stuff. Blockchain protocols may be good for some things . . . but they're too slow, too expensive, and too public. Personal data stores have to be fast, cheap, and private."[17] Berners-Lee came to this conclusion after a decade of attempting to crack the code of this intractable system.

Promise Unfulfilled: Implausible Revenue Models

Web3 social media isn't new. Even though some speak of Web3 social media in futuristic terms, it's actually been around for several years. A plethora of social media platforms already provide Web3 alternatives to the mainstream, centralized giants of Web2. Yet none have achieved mainstream success, and none have proven to have viable business models.

Web3's revenue relies extensively on *tokenomics* (token economics) baked into a company's platform. The cryptocurrency tokens used are most commonly Ethereum-related. Once a "coin" (another word for token or crypto) is selected, it is distributed to both the site's owners and its users. The number of tokens provided to the site's owners, employees, and users increases in tandem with user growth and engagement. Greater usage of the token/coin is expected to drive up its value. It's quite a plan . . .

Steemit was the first Web3 social network, emerging in 2016.[18] Somewhat of a combo between Facebook and Reddit, Steemit is a decentralized platform built on its own Steem blockchain. It enables users to earn cryptocurrency rewards for content creation and curation. Soon after Steemit's launch, other Web3 social media platforms like Minds (built on the Ethereum blockchain as a Web3 alternative to Facebook) and Voice (on the EOS blockchain) came on to the scene. More Web3 alternatives continue to proliferate, including BitClout, an alternative to Twitter/X; Dtube, PeerTube, and Odysee, alternatives to YouTube; Towns and Commonwealth, group chat apps competing with Discord; Mirror, a publishing alternative to Medium or Substack; and Audius, a network competing with Spotify. Still others are popping up.

Web3 networks have received massive funding. According to Crunchbase, blockchain-based startups raised a total of over $33 billion in venture funding between 2022 and 2023.[19] Despite geysers of funds and backing by major Silicon Valley venture capital, attempts to date have failed to gain mainstream traction. So why does the hype continue? Because Web3's tokenomics promises remain intoxicating.

Let's take a quick look at Web3's intended revenue models. Remember those crypto rewards we talked about? The owners and venture capitalists behind Web3 companies have been known to hold a significant percentage of the network's available native tokens for themselves.

Here's an eye-opening example: Compound and Uniswap are Web3 trading platforms funded by Andreesen Horowitz. As reported by the *New York Times*, "More than 95 percent of the coins that are used for governance on those two platforms are owned by just 1 percent of token holders."[20]

Unlike Web2, social entities in Web3 often charge their users transaction fees (with markup), also known as "gas fees," for completing certain actions on their platform. These fees are due to the built-in expenses associated with cryptocurrency usage. These fees can occur when a user sends or receives cryptocurrencies or NFTs on the platform, or if they engage in other kinds of transactions on the platform, such as getting upgrades and add-ons or making purchases. These transaction fees do two main things: They reward the crypto miners and validators who contribute to the blockchain, and they generate revenue for the platform.

Changpeng Zhao, the discredited former CEO of Binance, one of the largest cryptocurrency exchange platforms, told *TechCrunch* (prior to his downfall) that Binance "generates 90% of its revenue from transaction fees."[21] We don't need to delve into Binance's law-breaking activities. The point here is that at scale, there will be a significant number of transactions on a Web3 network. This traffic pumps up the value of its tokens, making its owners and investors rich. That's the master plan.

Here's the catch: Basing a network's value primarily on cryptocurrency is extremely risky. While many of crypto's early adopters have gotten rich, countless others investing in subsequent years have lost their shirts (the gamble is multifold—picking the right coin and riding it out). Placing this bet puts a network at risk due to this unpredictability, yet it remains alluring.

Another chink in the armor—cryptocurrency is in the crosshairs of governments worldwide. As of 2024, nine countries have banned cryptocurrency entirely,[22] and several others severely restrict its usage. The banning of crypto by more countries could potentially undermine Web3's entire foundation and plans for global social networking. Whoops, out goes the currency, there goes the revenue model.

Some Web3 social networks recycle greenback revenue features from Web2 such as paid ads (using either fiat or crypto payments), paid boosting of posts, extra feature enhancements, and premium memberships. However, in Web3, those revenue streams are bit players. Web3's chief revenue target typically is accruing profits from crypto.

Here's the core revenue roadblock: Web3 social media's failure to generate significant revenue is rooted in its lack of appeal to mainstream users. For its bundle of revenue schematics to work, critical mass must be generated, likely in the tens of millions to hundreds of millions of users.

Such numbers are not predictable at this juncture. As noted in a report by the market research company Forrester Research, "Many Web3 projects on offer run very well on PowerPoint"[23] but don't hold up in practice. The world doesn't appear ready to adopt Web3 at any level of real scale.

Blockchain Is Useful, but Not Yet for Social Media

There are plenty of potential benefits to blockchain, particularly in applications where multiple parties are involved. Everyone needs to see the same data, and proof is required that the data hasn't been tampered with. But currently, blockchain still is too cumbersome for the complexities of personal social networking.

Here's a positive example. UNICEF, the international charity supporting impoverished children, is currently exploring blockchain technology using digital smart contracts to distribute funds to schools in need. Its intention is to facilitate "new systems of trust and exchange on which users can send value directly from one party to another without the need for intermediaries."[24] Donors get transparency while distributing their donations to those in need across various countries and jurisdictions where there are varying currencies, laws, and regulations.

Blockchain is also useful in several applications in healthcare, including securing patient data and managing the pharmaceutical supply chain. Other uses being explored (though yet to be proven successful) include supply chains for items like fair trade coffee,[25] sustainably sourced palm oil,[26] and conflict-free diamonds.[27] This tech can potentially help customers trace and verify the sourcing of the items they purchase.[28]

Blockchain may be a workable solution for a spectrum of applications in finance, healthcare, charity, and supply chains. But it is mismatched for the assignment of personal social networking. Stephen Diehl, a software entrepreneur and Web3 critic, described the loss of computing power by switching from centralized platforms to decentralization: "We've gone from the world of abundance in [centralized] cloud computing where the cost of compute time per person was nearly at post-scarcity levels, to the reverse of trying to enforce artificial scarcity on the most abundant resource humanity has ever created. This is regression, not progress."[29]

Can Web3's slow and convoluted infrastructure support complex, full-featured, personal social media with hundreds of millions of users? Today, it's not up to the task.

6

Meet Your New Boss, Same as the Old Boss

Silicon Valley insiders and wealthy venture capitalists are clearly seeing dollar signs in Web3. The big boys are marching in. Indeed, the Web3 space appears to be led by an army of billionaires. A couple of examples: Jack Dorsey[1] (Twitter/X's original owner) has backed Nostr and Bluesky, two decentralized Twitter/X alternatives. (He's since parted ways with the latter.) Venture capital all-star Andreessen Horowitz is backing several Web3 sites.[2]

Here's the problem: A handful of venture capitalists (VCs) and tech investors already hold too much power over Web3's development, which gives average users the short end of the stick. VCs are buying up the tokens of new blockchains at early stages through private sales. So, it's much like traditional equity investments, just with investors owning significant slices of crypto tokens instead of shares. Later, the tokens are distributed more broadly to the public but at a far higher price tag. These investors stand to profit tremendously by Web3's adoption. No wonder they're such staunch believers.

In a system that's "democratized" via token ownership, regular users of Web3 are put at an extreme disadvantage. In the Web3 world, where tokens give you more power to control a platform, you'll now have governance driven and dominated by VCs and a handful of elite Silicon Valley insiders who are flush with tokens.

Coinbase noted in a blog post that this uneven system grants "majority control to a relatively small subset of people. As a result, the protocol is effectively more of a plutocracy than democracy."[3]

The same venture firms that funded and profited from Web2 surveillance giants like Meta and Twitter/X are the ones funding and hyping Web3. The irony is palpable.

Of course, Zuckerberg and Meta were not going to sit back and let the decentralized marketplace exist without shoehorning in. In 2023, Meta entered the fray with its Twitter/X clone, Threads,[4] built on Mastodon's decentralized protocol, ActivityPub. The app is a simple platform for influencers and regular individuals to share text-based messages with their followers. Apart from Threads, there is no plan to move Facebook to decentralization or Web3. This is telling.

According to Apple's App Privacy Details, Threads collects a plethora of your data including locations, contacts, purchases, browsing history, search history, financial info, health and fitness, and more.[5] This all makes sense—if you're Meta—masquerading while the data pipeline flows.

Vendors, too, who are involved in Web3 are also already showing monopolistic tendencies. It's the great arc of Web2 all over again. Behind the backdrop of this new promised land, most players in Web3 are using only one or two companies for their important services. For example, Infura and Alchemy are the new big boys for data retrieval in Web3, and Opensea is the main squeeze for nonfungible token metadata.[6] Holy smokes, Batman, Web3 is centralizing even faster than Web2 (which at least had a healthy variety of competing sites before Big Tech's consolidation took over). This market domination undermines the entire principle of decentralization.

In a blog by Martha Bennett, VP, principal analyst at Forrester Research, quoted in the *New York Times*, she said, "At this point, Web 3.0 is exhibiting the same monopoly-building, rent-seeking and value-extracting tendencies that its proponents decry about Web 2.0."[7]

By all appearances, the most likely outcome of Web3 is that the wealth and power of Web2 will merely shift from one set of tech billionaires to the next (and often to the very same billionaires as before). We appear to be running straight into the same smokescreen of Web2 Big Tech. Behind the smoke are not the arms of a new messiah but many of the people who created the Big Tech monster in the first place. *Meet your new boss, same as the old boss.*

7

Is There a Better Way?

I'm a capitalist and a fervent believer in its fundamental economic premises. We have freedom to imagine, invent, and improve, to delight and serve people by providing a product or service that enhances their lives, honorably, with trust and respect. This is the elixir that drives me. In my mind, this is capitalism's foundation. And it is certainly fair and just to reward entrepreneurs and all businesspeople who follow this simple recipe. There is no cakewalk to success—and I love the risk-taking entrepreneurial spirit that capitalism fosters.

Yet, somewhere along the line, capitalism lost its luster, misplaced its meaning, and has even become maligned and criticized in some social circles. The larger the enterprise, the greater the scorn. In a 2021 Gallup Poll, a mere 18% of Americans had confidence in big business,[1] and just one in three Americans had a positive view of Big Tech.[2] Hardly conducive to warm, fuzzy feelings.

Clearly, it's time for a transformation and an ascension. I call it "Restoration Networking" (also referenced in the book as "Restoration Networks"). This new path for social media protects, respects, and empowers the regular users most of us are and the aspiring and successful content creators among us, all while rejecting surveillance, targeting, and manipulation as revenue sources. How can we pull this off? By shifting to the rapidly proliferating, ethical, and fruitful economic paradigm called Conscious Capitalism.

(Note to skeptics: In the wake of Sam Bankman-Fried and his misappropriation of "Effective Altruism," we are correct to be wary of catchy buzzwords. But Conscious Capitalism is tried and true, with plenty of companies that have successfully adopted its principles. The proof is in the pudding.)

What Is Conscious Capitalism?

I've had extraordinary advisors in my social media odyssey. One of the most invaluable is my good friend Raj Sisodia, cofounder of the worldwide Conscious Capitalism movement. Raj wrote the bestselling book *Conscious Capitalism* with Whole Foods cofounder John Mackey.[3] Like all of us, Raj is aghast at capitalism's current distortion.

Raj has endeavored to bring forth an equitable future that fulfills the Conscious Capitalism credo: "We believe that business is good because it creates value, it is ethical because it is based on voluntary exchange, it is noble because it can elevate our existence and it is heroic because it lifts people out of poverty and creates prosperity."[4] To put it simply, Conscious Capitalism is doing well by doing good.

The philosophy of Conscious Capitalism combines profits and social responsibility. The *Harvard Business Review* has shown that companies practicing this modus operandi perform 10 times better than their peers.[5]

One of the core tenets of Conscious Capitalism is considering the best interests of everyone affected by the business. *Everyone*. In the case of Restoration Networking, that means a healthy alignment of revenue-sharing with all of us as social media users. This is not a Kumbaya sing-along. It's practical and empowering.

Of course, like any significant movement, Conscious Capitalism has its detractors. Some believe that it redistributes too much power away from business owners and investors. Others believe it doesn't redistribute enough. It's a fine balance to be sure, and its success speaks for itself.

There are many examples of major companies succeeding today with Conscious Capitalism. One of the most well known is Whole Foods. Yes, it was perhaps an even better global Conscious Capitalist citizen prior to being acquired by Amazon. Nonetheless, it still maintains several important tenets of its prior self. Notably, the company works with the Global Animal Partnership to certify its producers follow the 5-Step Animal Welfare Rating standards. It also forbids the use of artificial colors, flavors, and high-fructose corn syrup in any products for sale in its stores.

We can also look to the blossoming Benefit Corporations (B Corps) movement as another example of thriving Conscious Capitalism. B Corps support sustainable and responsible business practices, considering the impact of their actions on a variety of stakeholders, including employees, customers, their community, and the environment. Successful B Corps include REI, Ben & Jerry's, ButcherBox, Toms, Danone North America, Vital Farms, Clean Choice Energy, Warby Parker, and Patagonia. Each of these companies is highly successful and has excellent brand loyalty.

Perhaps you're a Patagonia fan. The company has reaped the benefits of boosted sales growth and strong brand loyalty due to its Conscious Capitalist principles. In fact, Patagonia's revenue has quadrupled over the past decade. In the words of former CEO Rose Marcario, "Doing good work for the planet creates new markets and makes [us] more money."[6]

Salesforce, Trader Joe's, The Container Store, and many others, while not B Corps, operate as fully declared Conscious Capitalist enterprises. They demonstrate that the world is ready to spend its dollars with companies that embrace business practices making the world better and healthier.

The tech space is already proving that companies can do well by doing good too—even if they're not anointed "Conscious Capitalist."

Take the search engine DuckDuckGo, for example. You might have used it, seen its billboards, or heard its ads. It competes with the Goliath known as Google via several differentiations—it doesn't track users, personalize results, or send your information to advertisers. Has this resonated with users? Absolutely. In 2023, the company reportedly had 100 million users.[7]

Then there's Apple, one of the largest companies in the world with a market cap in 2024 of over $2.8 trillion (that's trillion with a *t*!).[8] Apple makes products that many of us love, and—notwithstanding significant issues endemic to Big Tech, including anticompetitive practices—it successfully markets itself as a privacy-focused company that rejects Surveillance Capitalism. Maybe you've seen one of Apple's billboards while driving—many of them say in giant letters "Privacy. That's iPhone."

Conscious Capitalism for Restoration Networking

The bottom line is that Conscious Capitalism (i.e., respecting customers, vendors, users, and employees as valued stakeholders), while highly effective on an uplifting, feel-good level, can also generate tremendous profits for practicing companies. It's time for personal social networking to do the same. The next sections explain how.

Users Are Partners

Restoration Networking shatters the paradigm of conventional social media monetization. Its economic model serves and embraces users as a specific shareholder class of the company. Imagine this: We're valued as partners instead of peons. How's that for a reset?

Several Conscious Capitalism companies have already implemented revenue-sharing models to great success. Many B Corps also incorporate profit sharing into their business model, sharing a portion of profits with their employees and partners. For example, sustainable fashion brand Eileen Fisher is 40% employee-owned and shares a portion of its profits with employees.[9] Patagonia shares 1% of its sales with environmental nonprofit organizations,[10] in addition to profit sharing with its employees. And organic chocolate maker Alter Eco shares a portion of its profits with the farmers who grow its products and even funds their transition to regenerative farming.[11]

Profit Sharing via User Awards

Restoration Networks logically bake in profit sharing with their users via a User Awards program. After all, user activity and engagement are really the engines that drive revenue. Through the program, regular users earn a distributed share of the revenue generated by meeting certain criteria. In this paradigm, you're not just a user of the social media platform—you're a stakeholder. Criteria can include a user's participation, followers, content creation, and overall site usage.

A certain percentage of the company's net profits—let's say 20%—is evenly distributed among all users who meet these qualifications. This includes revenue and profit sharing with:

- Active users
- Premium subscribers
- Content creators
- Influencers
- Page owners

There are additional ways to earn User Awards, Including bringing new active users to the platform or owning a page or group that surpasses a certain threshold of active followers or members. Users can also submit helpful bots or suggest updates to the site's open-source code. These User Awards support the platform's core philosophy of rewarding users who contribute to its financial success.

After all, if we as users are helping the platform make money, it only makes sense that we earn a fair slice of the revenue pie, right? It's win-win, pure and simple.

Protecting and Respecting User Data

Right now, pretty much everybody on the Web has their data stored in a centrally controlled data infrastructure. There it is sliced, diced, and parsed among massive data aggregating companies and Surveillance Capitalist corporations like Meta, Google, and Amazon. Your data circulates through these arteries to advertisers, marketers, governments, and whomever wants it.

There's a better way in perfect harmony with Conscious Capitalism's principles. Aligning in partnership with its users, Restoration Networks by design will have two choices engineered to protect their data. One is a centralized data storage repository provided by the network that does not share user data with advertisers, marketers, or any other third parties except for those needed for essential functionality.

The second is the creation of Web inventor and privacy advocate Tim Berners-Lee. He has worked hard in recent years to revolutionize how we store, manage, and share our data and content—via Pods (Personal Online Data stores), which are part of his Solid protocol initiative.[12] Imagine your personal data (your info, photos, documents, stories, videos, contacts, etc.) as valuable items that you keep in a digital safe. This safe (your Pod) is under your control, and only you have the key to open it. You decide what goes in and what goes out for others to see.

The centralized user data bank of Restoration Networks can be engineered to be compatible with Pods and their underlying Solid protocol. In this way, users have the option to upload and download their data/content to and from the network via their private Pod at any time. Optionally, Pods can automatically download your data, so you always have an up-to-date copy ready to port wherever you want. That's data storage and posting management with true privacy protections. R-E-S-P-E-C-T!

Since all your data is stored in your own personal Pod, you can conveniently log in with a single ID to apps, sites, and services, anywhere across the Web. Thus you can enjoy the benefits of "single sign-on," which we covered earlier in the promises of Web3. As an additional important bonus, Pods do not use blockchain, thus sidestepping its myriad drawbacks.

From Berners-Lee's site, you can choose which Pod provider you wish to store your data with, or you even can learn how to self-host your own Pod. As an added benefit, you can move your data between Pod providers whenever you want. This is a pretty sweet solution—with plenty of provider choices already. In fact, it's already catching on in Europe—Berners-Lee noted in 2024 that the Flanders region of Belgium has authorized Pods for all its citizens.[13] Other high-quality vendors are sure to roll out their solutions as well.

Giving all of us control over our own data is a worthy sentiment and initiative. This solution can work for a minority of tech-savvy users, but not for the majority of ordinary folks online. But it also saddles people with responsibilities they may not wish to have or aren't capable of fulfilling. That's the beauty of the approach of Restoration Networks—you can let the network host and manage your data for you, or you can handle your data via Pods or other emerging privatized solutions. The choice is yours.

Positive Inevitabilities in Conscious Capitalism

One of the foundational drivers of the capitalist startup ecosystem is the ability for founders and investors to have a profitable "exit." In the tech world, this exit is accomplished via an initial public offering (IPO) or, more commonly, an acquisition of the startup by a larger company.

Reddit's IPO in 2024 is an interesting example in which the company gave its most active users the option to purchase shares at the offering price before the general public was able to acquire them.[14] Of note, historically, these shares would have been reserved for insiders, such as Reddit's investors and other large financial entities.

Conscious Capitalist companies transact just like other profitable companies. So, you might be asking yourself—can a Restoration Network be sold or acquired by another company? Yes, it's inevitable that a successful company in any genre will have exit options that serve its founders and investors. That's what incentivizes many of the best minds and developers in the world to devote their careers to these challenging endeavors in the first place. We can look to Whole Foods' acquisition by Amazon, SolarCity's acquisition by Tesla, or WhatsApp's acquisition by Meta as examples.

But here's what makes Restoration Networking unique in this regard. By design, via their engineering with data interoperability, Restoration Networks provide users with the option to easily move, along with all their data, content, social graphs, contacts, and so on, to another network at any time. As a user, you are not married to any particular network. If an exit of your favorite network is unsavory, you can seamlessly migrate elsewhere.

Additionally, since Restoration Networks are built with privacy by design, they don't collect your personal data (beyond what's needed for you to log in to your account), unless you opt to share it for monetary rewards. Therefore, you don't need to worry that a sale or acquisition will mean that your personal info—your relationships, interests, political views, and others—will fall into the Surveillance Capitalism ecosystem. That information was never collected by the Restoration Network in the first place. Everybody wins.

8

Empowering You: Social Media User, Creator, Star

Who would have thought that millions of people would become the fuel driving economic engines of countless social media companies? Yet this is the new reality we're participating in, regardless of our roles. Today we all know someone—a friend, family member, classmate, or colleague—who is a current or aspiring content creator. Maybe you're a parent with a child, niece, or nephew who wants to be a social media star. Perhaps you have dreams of becoming one yourself. CNBC reported in 2023[1] that 57% of Gen Zers want to become social media influencers.

A 2022 study by Adobe suggested that over 300 million people worldwide consider themselves to be content creators. Equally remarkable is that over half of them have joined the creator economy just since 2020.[2]

The Slog to Stardom

There are numerous inspirational examples of people who went from complete obscurity to fame with zero experience or connections in the entertainment industry, to diligently building up their followings and skyrocketing into social media stardom. Look at Jimmy Donaldson, better known as "MrBeast." He started out as a 13-year-old kid making and uploading YouTube videos about video games from his bedroom in his parents' house. Today, he has over 250 million subscribers on

YouTube and reportedly earns over $80 million per year.[3] Not a bad chunk of change for a former video game enthusiast.

Or take a look at "That Chick Angel,"[4] who was a relatively unknown rapper and performer. After she recorded the song "One Margarita," it went viral on TikTok and completely changed her life. Thanks to her song's social media success, she ended up making an official music video with none other than former supermodel Cindy Crawford.

While it looks easy to the uninitiated, being a professional content creator with a strong following requires years of painstaking work and dedication. Just like those who aspire to careers as professional athletes, only a fraction will earn enough to eke out a living and just a handful will create extraordinary wealth. MrBeast acknowledged this reality in a post on Twitter/X in 2024: "For every person like me that makes it, thousands don't."[5]

Despite their hard work, most creators earn next to nothing. According to a 2023 report by Linktree,[6] only 12% of full-time content creators earn $50,000 or more in annual revenue. Meanwhile, 46% of full-time creators earn less than $1,000 per year. When it comes to part-time creators (those who create content as a side hustle or passion project rather than their full-time career), only 3% earn more than $50,000 per year, and 68% earn less than $1,000 per year.

Sharing the Spoils

How is it possible that so many creators, including those who have tens of thousands or even hundreds of thousands of followers, so often struggle to make enough money to survive? The problem is that the spoils of their hard work are not being shared properly. The Big Tech social media companies are reaping nearly all the rewards while leaving creators—the breadwinners of their platforms—to nibble at the crumbs. Yet content creators are the backbones of any social network. The followers they attract and the content they create keep regular users returning, engaged, and entertained, which provides enormous value to social networks.

How can we change the economic model so that we're all empowered? We all have skin in this game.

Right now, content creators (and all of us) are subject to the fickle whims of Big Tech companies, which are the true owners of our content, our audiences, and our destinies on social media. Big Tech's opaque algorithms muzzle our reach, reducing our voices from powerful megaphones to faint whispers. For example, the average Facebook post only reaches 5% of its page followers.[7] Perhaps you've noticed that, on Facebook or Instagram, you see fewer and fewer posts from your friends or the people and pages you follow. Instead, you see an endless parade of ads and promoted content.

That's because Big Tech platforms are in the advertising business. They want creators to fork over their hard-earned cash just for the privilege of reaching their own followers, whom they've worked so diligently to cultivate over the years. Does that sound fair to you? Me neither.

Restoration Networking champions win-win relationships between a creator and the social networks they post on. For the creator inside you, here's what the future can look like.

100% Follower Reach

Restoration Networks vaporize the Big Tech algorithms and unfasten the muzzles on all of us—creators and regular users alike. Using Restoration Networks, 100% of our posts reach 100% of our followers and connections. It's our microphone, so let's get blabbing without Big Brother.

True Portability/Interoperability

Web2 social giants are walled-in silos. If you want to move from site to site, you can't. Web3 boasts about its data portability benefits, but as we saw in Chapter 5, Web3 is mired with so many drawbacks that currently the juice isn't worth the squeeze.

Restoration Networking provides the regular users and creators among us with true portability in moving our content, contacts, and fans/followers from site to site, which guarantees we all have true freedom while supporting fair competition among social media companies. This is called "data interoperability."[8] Single sign-on makes this interoperability possible by allowing you to easily log in to any Restoration Network using a single ID. Expanding on this, if a Restoration Network is connected to Pods, you'll have an additional universe of sites to seamlessly post to and engage with.

A Restoration Network can also be engineered so that user data is interoperable among protocols outside the Restoration Networking ecosystem. For instance, user data could be interoperable between a Restoration Network and applications using the Solid protocol we discussed in Chapter 7, or with ActivityPub,[9] the social networking protocol used by sites such as Mastodon, which we mentioned in Chapter 5, as well as Threads, PeerTube, and WordPress. As a caveat, this structure will not be the standard default due to privacy concerns. Instead, you the user will have the choice to opt in, allowing your data to be interoperable with sites outside of Restoration Networks if you'd like.

To encourage a robust creator economy and free market, Restoration Networking supports new legislative regulations to limit content creator mobility lockups, thereby eliminating anticompetitive constraints. (See Appendix A.)

True Deletion

You might not realize this, but most Big Tech social media companies keep displaying the content you've created on their platforms, even after you delete your account. These sites apparently think they are the owners of your content, not you. Restoration Networks disagree. If you decide to abandon one site for another, all your content from the former site will be deleted.

Feed Curator: Newsfeed Customization

Restoration Networking empowers you, the user, by giving you robust options to design your own newsfeeds. With the help of AI tools available within the platform (don't worry, we'll dive deeper into AI in Chapter 15), you can customize and curate your newsfeeds for whatever topics or hobbies you like. For instance, you could create a feed that collects and spotlights content related to your passions. Perhaps it is content related to sports, food, photography, fitness, fashion, movies, or others. This is a healthy, enriching, and delightful use of AI.

Built-in AI tools enable you to become a newsfeed architect—no engineering/coding skills are required. You also select the AI filters to keep out content you deem distasteful. Optionally, you can then choose to make your curated feeds available for your friends, family, or anyone else on the platform to follow. Users can upvote their favorite user-curated feeds. You can even be monetarily rewarded by the platform for creating feeds that reach a certain threshold of popularity. This innovative feature makes "Feed Curator" a whole new category of influencer on Restoration Networks.

Show Me the Money

In the Conscious Capitalism spirit, users and creators among us are rewarded for our contributions. We all share in the revenue pie. Remember the User Awards we talked about in Chapter 7? Restoration Networks use a simple formula to determine a fair User Awards payout. It's based on the percentage of site activity that each of us generates and participates in. There will be minimum thresholds. Users with minimal onsite activity won't qualify, as otherwise the company will be logjammed with user verification paperwork, documentation, 1099 issuances, and the like.

This meritocratic system has two key benefits—one, it rewards us fairly for our contributions to the platform, and two, it provides an incentive to those of us who want to grow our audiences and enjoy posting engaging content.

Here's an enticing example. Let's say a company has a net revenue of $100 million for the year, and it pays 20% of that to its users. If you're a creator generating 0.001% of the site's activity (as defined in Chapter 7), you would earn $20,000. Here's the asterisk (*): The site in tandem with its users and AI tools will monitor and weed out attempts to "game" its reward system. (More on AI later.)

In line with Restoration Networking's pro-democracy principles, decisions regarding User Award distributions will be determined by the User Advisory Board, which we'll dive into in Chapter 10. It's all about creating a fair balance of revenue for the company's owners, stakeholders, and users.

So, in Restoration Networking, users share in the monetization of the platforms they're active on. Show me the money!

9

Do I Need a Social Media Wallet: Is There Money in It?

You might have heard about social media wallets, particularly in tandem with the crypto movement. We've heard big companies talking about them, and early crypto adopters seem to have them. Yet they still seem shrouded. What do they entail, and is there anything in there for you? Currently, most social media wallets live within their host app. Each user gets one. They let you receive, store, send, and trade digital assets, such as cryptocurrencies and nonfungible tokens (NFTs). Importantly, they also store the keys for single sign-on. Web3 is where they're typically deployed. As an example, early Web3 social networks like Minds and Steemit have them. Their wallets are key components of the tokenized reward systems on those sites.

Restoration Networking Wallets

If you like to gamble or play the lottery, cryptocurrencies are perfect for you. They're just not suitable for everyday transactions on a social media platform. Restoration Networks recognize these facts by sidestepping the crypto craze entirely. Instead, Restoration Networks have your back with greenbacks.

Forget Crypto as the Centerpiece

Here's the most significant differentiator. All payments sent to your wallet by the company are made with fiat currency. That means real, cold hard cash, such as the US dollar or whichever national currency you use in your country. This is a major break from the Web3 model, which is entirely based on cryptocurrency.

Besides their schematic voodoo, the cryptocurrencies inside these wallets are extremely volatile—their prices fluctuate wildly.[1] They're about as a stable as a kayak in the middle of the sea during a typhoon. To be clear, I'm a crypto investor, but I don't see it as the elixir for a Restoration Network. Fiat takes that role, front and center.

As an important caveat, while fiat currency is the default, Restoration Networks can also choose to include certain cryptocurrencies in the future if/as they become dominant, predictable, stable, and easily liquid.

Cashing Out

The best part? Restoration Networks allow you to cash out your credits on the platform for real money (transferred to your bank, PayPal, Stripe, etc.) anytime you'd like. This is a major departure from versions of this system such as Reddit gold, which you might be familiar with. On Reddit, you can punch in your credit card to buy Reddit gold, which you can use to buy enhanced features or digital gifts for other users. But you can never cash out—once you put your money into Reddit's system, it stays there forever.

So, users like you are going to be earning money from other users and from the Restoration Networking company via User Awards. What's the easiest way we can facilitate this?

Your Restoration Networking Wallet

To make it simple, upon joining a Restoration Network, all of us will automatically receive a wallet. All User Awards, as well as additional payments you receive, are deposited into your wallet. To keep it simple, once you've got your wallet, you can use it across Restoration Networks.

Restoration Networking rewards all active and qualifying users via their wallets, not just celebrities and influencers, with a share of the company's annual profits. That's a revolutionary upgrade from previous generations of social media. In this paradigm, we're all stakeholders.

This stakeholder status raises the question: How can you as a user make your voice heard regarding the platform's decisions and directions?

10

How Do We Overcome C-Suite Tyranny?

Executives like to consolidate their control of the companies they reside within. We've all seen it. Many of us are guilty as charged. Capitalism is not fundamentally democratic, especially in the entrepreneurial genre—it's rather autocratic with founders and investors vying for domination. Societies function similarly when unshackled from the constraints of civility. Just as we all require speed limits and laws to buckle up, Restoration Networking ups the game with checks and balances intended to preempt power grabs and Surveillance Capitalists from roaring back.

As I mentioned earlier, I love capitalism, and I abhor overregulation. Yet with just a few simple self-regulatory structures, Restoration Networking ascends the Web1 and Web2 models in which all power resides under a single executive—think Mark Zuckerberg who rules as emperor over billions of voiceless users. At the same time, Restoration Networks carefully avoid the significant pitfalls of Web3. Here are a few ways that Restoration Networks promote true democracy in their social media experience. You can participate in these; we all can.

User Advisory Board

To ensure fairness and that all users have a strong voice, each Restoration Network has its own User Advisory Board. This board is composed of users nominated by their peers, tasked with suggesting

User Award distributions, providing feedback, and suggesting new features and policies.

Several major companies have established user/customer advisory boards to obtain valuable feedback and insights from their users/customers. For example, Microsoft's customer advisory boards offer feedback on product development,[1] roadmaps, and user experiences. The boards also have access to early product builds and the ability to provide feedback to product teams. Airbnb has a community advisory board comprised of hosts from various countries.[2] These hosts periodically meet with Airbnb leadership to discuss concerns and opportunities within the host community.

A Restoration Network's small but representative User Advisory Board ensures that regular users play an active role in shaping its future, which preempts decision-making gridlock. That gridlock occurs in overdemocratized entities with too many cooks in the kitchen. Instead, the network maintains a dynamic equilibrium, allowing swift and effective management while empowering and remunerating users. The interests of owners, users, and stakeholders are all in play. The objective is a win-win-win balance of power. (For further details about the User Advisory Board and how it works, check out Appendix B.)

The Restoration Networking Institute

How do we hold Restoration Networking companies accountable to the movement's principles? What are the best ways to prevent them from simply paying lip service? Once again, we can look to the established standards of the regenerative farming and renewable energy movements.

Here are two pretty awesome examples: The Rodale Institute and the Center for Resource Solutions. Both are nonprofit organizations, with about a dozen members on their boards. They're composed of experts in relevant fields. Funding comes from a combination of grants, donations, consulting services, and collaborations with business partners in their industries. The Restoration Networking Institute can follow a similar path.

The Rodale Institute, which first coined the term "regenerative farming," has similarly developed an official certification, "Regenerative

Organic Certified," defining how a farm may operate in compliance with regenerative principles.[3]

In the renewable energy sector, Green-e is an industry-trusted, third-party certification program from the Center for Resource Solutions that reviews renewable energy and carbon offset products.[4]

Similarly, Restoration Networking companies can be certified via an independent nonprofit entity comprised of diverse and respected leaders of technology and ethics. For our purposes, let's call it the Restoration Networking Institute.

Bringing this Institute to life, the collective of Restoration Networks can, as they permeate, fund the Institute and establish its operating principles. Once extant, it is tasked with delineating criteria for its raison d'être—the Institute's "Certified Restoration Network" verification. This certification ensures that anointed networks/apps have met the ethical and business practice standards and principles of Restoration Networking. A certification from the Institute signifies that a Restoration Network meets the gold standard of the movement.

As an aside, perhaps you've heard of Meta's Oversight Board. This independent advisory group was established a few years ago to help Meta pass the buck on difficult moderation decisions. Notably, its first major task was to determine whether President Trump should be allowed back on Facebook after he was temporarily banned in 2021. The board's big decision? It declined to decide and instead passed the buck right back to Mark Zuckerberg. So much for helpful "oversight"![5] Fast forward to 2024: The Oversight Board made better news when it called Meta's policy to tackle deepfakes "incoherent."[6]

The Restoration Networking Institute is entirely distinct from any single network—it serves to independently validate whether companies are upholding the movement's principles, not to volley back and forth about moderation or policy decisions. (For more details, head to Appendix B.)

Changing of the Guard

As we've learned from Big Ag and Big Energy, real change comes from outsiders leading the charge. Transformation won't come from Web2 social media giants like Meta, Google, or Twitter/X. Nor will it come from the same VC investors who got wealthy from those Web2 giants now funding the Web3 movement. We need new people, new platforms, new investors, and new paradigms to lead us toward a healthy social media future. Is that you?

■ ■ ■

What about all the other elephants in the room—privacy, anonymity, protecting kids, mental health, AI, bots and trolls, free speech, and others? We'll tackle them all. First, let's start with privacy and anonymity.

11

In the Crosshairs: Privacy and Anonymity

Let's talk about privacy. Believing in my right to privacy comes from my gut and my experience as an American. I think of privacy as an innate right as a human being. Here's how I see it: I am allowed to think what I think, act as I choose, and associate as I wish, without being unnecessarily spied on or otherwise targeted, as long as I am law-abiding.

In the larger context, the right to privacy ensures that individuals are free to express themselves, make decisions about their own lives, and live without fear of discrimination or persecution based on their identity or beliefs.

A few years back, after an outbreak of revenge porn Down Under, Facebook asked its Australian users to upload naked photos of themselves.[1] Their hare-brained plan was to scan the photos received into a special database for image matching on the site, thereby preventing bad actors from sharing them. You better get to a tanning salon.

Imagine the gold mine for hackers! Meta, a company infamous for its massive data leaks[2] and privacy breaches, would have a database of nude photos of millions of its users. Fortunately, this experiment quickly failed due to the common sense of the requested users who refused such a comically absurd proposition. This vignette illustrates the challenges we face with Big Tech that is out of touch with the fundamentals of privacy and anonymity.

You know how we like to muse about the "good old days"—those simpler times in the past? It's not just nostalgia over lost privacy; things truly have gone awry. Thirty years ago, privacy was a given in the United States. Sure, neighbors could be nosy, but folks generally minded their own business. They weren't government agents paid to spy on everyone else. This situation starkly contrasted with nondemocratic countries and their citizens who routinely reported the conversations and movements of others, including friends and family members, to the authorities.

Now the world is different. The old ideals of privacy have been shredded. We don't need our neighbors to be the secret police. Our personal computers, cellphones, cell towers, smart TVs, Alexa, and cameras everywhere (including our own) efficiently spy on us better than any human could. Yet, paradoxically, we're also able to anonymously bully others, post and share outrageous content, and otherwise act online as derelict global citizens.

Today, privacy and anonymity are both endangered. Rightfully so? If we lose the privacy of our thoughts and circumstances (never mind our nude photos), what are the risks—and what are the benefits? Who will get to decide their fate—another Mark Zuckerberg, or the government? How do we balance the polarized objectives of protecting a user's privacy while holding them accountable for actions violating the law or a site's Terms of Service? How do we eliminate anonymously boosted content and the manipulation of our opinions/purchase decisions? How do we protect the rights of free speech and nurture civil discourse?

The answers to these questions have huge implications for the future of humankind along with the reach and influence of its governing structures. This is the gladiatorial clash between privacy and anonymity.

What Does Privacy Really Mean?

What is privacy, anyhow, and why is it so important? In broad terms, privacy is the right to be left alone, without having to worry about someone intruding into your life and personal affairs. Of course, the situation gets complicated in our ubiquitous digital age. Let's add some color to this definition:

- The International Association of Privacy Professionals describes privacy as "the right to have some control over how your personal information is collected and used."[3]

- Your personal information includes any data that can identify you as an individual: your name, address, date of birth, Social Security number, biometrics, and others.
- It also includes potentially sensitive details about you, such as your race, ethnicity, religious views, sexuality, mental well-being, finances, medical records, political views, opinions, personal relationships, and virtually everything about you.
- Privacy means that you have the right to decide what about you is revealed and what is not.
- It means you have the right to know what personal information is being collected about you, why it is being collected, and who it is being shared with.
- Privacy also means that you have the right to give or withhold consent for the collection, use, and disclosure of your personal info, and you have the right to withdraw your consent at any time.

Notable Quotes by Privacy Champions

While Surveillance Capitalists, authoritarian governments, and countless others try to downplay the importance of privacy, many luminaries still righteously defend its virtue. Here are a few notable quotes from contemporary privacy advocates:

> Privacy, a core American value, is not a partisan thing. Democrats fight for it and Republicans fight for it, too.
>
> **—Tim Berners-Lee, inventor of the World Wide Web**[4]

> We believe that people have a fundamental right to privacy. The American people demand it, the Constitution demands it, morality demands it.
>
> **—Tim Cook, CEO of Apple**[5]

> Privacy knows no borders: we have to protect privacy globally or we protect it nowhere.
>
> **—Dr. Ann Cavoukian, former Privacy Commissioner of Ontario**[6]

> Privacy is not about having something to hide, it's about the right to control what you want to keep to yourself.
>
> **—Edward Snowden, former NSA contractor and whistleblower**[7]

Is Privacy Dead? Notable Quotes by Privacy Cynics

On the flip side, some people, especially those in the highly profitable business of Surveillance Capitalism, have a fundamentally different view. Perhaps in an attempt to weaken public demand for privacy, these folks argue that privacy is already deceased, and won't be resurrected anytime soon. In their view, we should all just give up the ghost and accept that our privacy is kaput. Along with the tech titans, some tech journalists have embraced this perspective as well.

Let's read their words, particularly chilling for anyone who has believed them and accepted privacy as a thing of the past—but without having any say in doing so:

> If you have something that you don't want anyone to know, maybe you shouldn't be doing it in the first place.
>
> **—Eric Schmidt, former CEO of Google**[8]

> People have really gotten comfortable not only sharing more information and different kinds, but more openly and with more people. That social norm is just something that has evolved over time.
>
> **—Mark Zuckerberg, CEO of Meta**[9]

> You have zero privacy anyway. Get over it.
>
> **—Scott McNealy, former CEO of Sun Microsystems**[10]

> Privacy is dead, and social media holds the smoking gun.
>
> **—Pete Cashmore, founder of Mashable**[11]

How Does Anonymity Differ from Privacy?

To put it simply, anonymity is when your identity is unknown. For instance, let's say you share a post on a public platform like a social network, but no one can tell it was you who posted it. That's anonymity.

Anonymity has the potential to let people express themselves more freely. Cloaked in secrecy, they can feel safeguarded from any possible pushback or discrimination. But anonymity has a dark side. It can be used to shield perpetrators of online abuse, bullying, hate, intimidation, and illegal activity. Read on—it's time for a remedy.

Privacy is about having control over your personal information. In other words, you decide who knows what about you. For instance, your neighbors know that you live across the street. They (probably) know your name. They see you come and go. It's the same thing at work. What you reveal about your personal life is up to you. The differentiator between anonymity and privacy is being known versus there being no awareness of your existence. For example, when you post anonymously, there's awareness of a person (or a bot/troll), but no awareness of *you* per se.

A Brief History of Privacy

There is a misconception about the concept of privacy: that it is explicitly protected by the Constitution of the United States. It's not. But the *right* to privacy has been established over the centuries in the Supreme Court's interpretation of the Constitution. These concepts have evolved, shaped by a variety of factors, including technological advancements, legal and regulatory developments, and cultural and societal norms. Here's a quick timeline to give us our bearings in the current privacy landscape.

The Enlightenment and the Birth of America

In the 17th and 18th centuries, Enlightenment thinkers such as John Locke and Jean-Jacques Rousseau began to articulate the concepts of individual rights to liberty and independence from the state and other powers. Their work helped pave the way for the individual rights to privacy that we cherish today. As Locke wrote in 1689, "The end of law is not to abolish or restrain, but to preserve and enlarge freedom,"[12] and as Rousseau wrote in 1762, "To renounce liberty is to renounce being a man, to surrender the rights of humanity."[13]

The Enlightenment inspired the authors of the US Constitution, which came into effect in 1789. Although it doesn't cover privacy explicitly, the Supreme Court has determined that the First, Third, Fourth, and Fifth Amendments of the Constitution do encompass the right to privacy.[14]

The Industrial Revolution

In the 19th century, the rise of mass production and urbanization led to increased surveillance of workers by employers and governments. Along with new technological advances such as photography, concerns arose about the potential misuse of personal information and the need for privacy protections. In the 1890 *Harvard Law Review*, Supreme Court Justice Louis Brandeis and attorney Samuel Warren penned "The Right to Privacy" (or "the right to be let alone").[15]

This remarkable article is recognized by many as America's seminal text advocating for our rights to privacy. Brandeis and Warren wrote that "instantaneous photographs and newspaper enterprise have invaded the sacred precincts of private and domestic life," and they proposed to "consider whether the existing law . . . can properly be invoked to protect the privacy of the individual."[16] *Just making sure this sinks in: 130+ years ago, photos on the fly and posts (in newspapers) were already giving people the creeps.*

World War II and the Cold War

In the mid-20th century, the rise of fascism and communism fostered an explosion of government surveillance and censorship of citizens.[17] This monitoring led to hearty debates in the West and worldwide about the need for protections against government abuses.

In 1948, representatives of the United Nations drafted the UN Declaration of Human Rights. Article 12 of this breakthrough declaration states: "No one shall be subjected to arbitrary interference with his privacy, family, home, or correspondence . . . Everyone has the right to the protection of the law against such interference or attacks."[18]

A year later, in 1949, George Orwell published his iconic dystopian novel *1984*, which illustrated the perils of government power, censorship, and surveillance. In Orwell's book, citizens of Oceania live with zero privacy. Their lives are constantly monitored by ubiquitous cameras and microphones. They're punished for any whiff of "wrongthink," thoughts or beliefs that question or contradict the ruling party. This novel inspired generations of readers worldwide, myself included, to advocate for the right to privacy. Again, what we're dealing with is not new; 75 years later, Orwell is as relevant as ever.

The Civil Rights Era and Fighting "The Man"

In the 1960s, the Civil Rights movement promoted privacy as a vital tool for protecting vulnerable and marginalized communities from discrimination. During this period, several Supreme Court cases established important precedents for the right to privacy in the United States. These cases continue to influence privacy law and policy today.

For instance, in the 1965 case of *Griswold v. Connecticut*,[19] the Supreme Court held that a state law prohibiting the use of contraceptives violated the right to privacy. Justice William O. Douglas pointed to the First, Third, Fourth, Fifth, and Ninth Amendments as foundational to the right to privacy, each iterating another aspect of our rights.

In the 1967 case of *Katz v. United States*, the Supreme Court determined that the Fourth Amendment protects people, not just places.[20] The Court ruled that a warrant is required for electronic surveillance of private conversations, even if the conversation takes place in a public phone booth.

It was in the Civil Rights era that the term "The Man" became popularized. "The Man" refers to whatever person or entity has the position of authority, representing an oppressive force against us as individuals. It's The Man who Rousseau, Locke, and Orwell stood up against. The Man always wants to spy on you. Protecting your privacy means opposing The Man. It's easy to spot The Man today—just open your Meta app. Big Tech corporations are listening and watching.

The Digital Age: An Unwritten Future for Privacy

In the late 20th and early 21st centuries, the invention of the Web and social media opened a Pandora's box. The rise of Surveillance Capitalism ushered in a worldwide erosion of privacy.

Tim Berners-Lee did not create the Web so that our privacy could be eradicated. Thankfully, Berners-Lee has been an outspoken privacy advocate.[21] With initiatives like Pods and Solid[22] (discussed in Chapter 7), he and like-minded luminaries are fighting to restore and advance the Web based on the original principles of his invention.

Other initiatives underway also promise privacy protection. Not all are in our best interests. Let's be cautious with initiatives like Web3, which can undermine this promise by placing our personal interactions on a permanent and public blockchain.

The future of our privacy remains unwritten. The battle continues, but I'm optimistic that we can save the digital age from the abyss of a *1984*-style dystopia. We can build a future where user privacy is restored, respected, and protected. And we must.

Laws Protecting Privacy

In tandem with the trailblazing court cases, several important privacy laws and regulations have been established over the years. They aim to protect individuals' privacy rights by ensuring that entities are held accountable for how they handle personal information.

The Health Insurance Portability and Accountability Act (HIPAA), passed in 1996, regulates the use and disclosure of personal health information by healthcare providers and insurance companies.[23] The Children's Online Privacy Protection Act (COPPA), passed in 1998,[24] requires websites and online services that collect personal information from children under the age of 13 to obtain parental consent and provide certain privacy protections. In Europe, the General Data Protection Regulation (GDPR), which went into effect in 2018,[25] is a comprehensive privacy law that regulates the collection, use, and disclosure of personal data by organizations operating in the European Union (EU).

More recently there's the American Privacy Rights Act (APRA), unveiled in 2024.[26] We'll see how that proposed act plays out.

While these and other laws take steps in the right direction, the reality is that regulations are famously slow in keeping up with technology. It's an unfair race between high-tech cheetahs and bureaucratic tortoises.

Are There Laws Protecting Anonymity?

There are no specific "anonymity laws" per se in the United States or worldwide that protect a person's right to anonymity in all situations. But there are laws and regulations that provide some level of protection for anonymity in certain contexts.

For instance, the Supreme Court has recognized that anonymity can be an important aspect of free speech enshrined by the First Amendment,[27] allowing individuals to express themselves without fear of retaliation. The 1995 Supreme Court ruling in *McIntyre v. Ohio Elections Commission* states: "Anonymity is a shield from the tyranny of the majority. . . . It thus exemplifies the purpose behind the Bill of Rights, and of the First Amendment in particular: to protect unpopular individuals from retaliation . . . at the hand of an intolerant society."[28]

There are also whistleblower laws, such as the Whistleblower Protection Act in the United States.[29] These critical laws protect the anonymity of whistleblowers who report illegal or unethical behavior. Vitally, they encourage whistleblowers to come forward and speak truth to power without fear of retaliation. In this context, anonymity protection is crucial.

Privacy's Stronghold: Encryption

America's founders declared independence from Britain in 1776 in part to stop authorities from intruding into their private lives. While our privacy has been significantly eroded in today's brave new digital world, there remains one last refuge: end-to-end encrypted messaging. Offered by services like Signal, Telegram, iMessage, and WhatsApp (either by default or via upgrades), end-to-end encryption is the final stronghold of personal privacy.

Collectively used by billions of people, along with companies, health organizations, and governments worldwide,[30] encryption enables messages, images, videos, and other content to be viewed solely by the sender and the recipient. Even the service providers cannot access that content. Private, encrypted messaging is critical. It safeguards communications for journalists and their sources; doctors and their patients; governments; banks; whistleblowers; human rights activists; political dissidents; and all of us who wish to protect against surveillance and data breaches.

It is natural to have aspects of our lives that we prefer to keep private. There are things you might confide to your spouse, doctor, lawyer, or therapist that you would never want revealed to the world.

Even if most of us are not journalists, activists, or political dissidents risking the wrath of government authorities, we still all benefit from the work those people do. It's essential that we collectively protect their efforts and their ability to communicate safely in private.

It's Complicated

Some policymakers in the United States, Europe, and around the world are working overtime to conquer this final frontier of privacy. We are witnessing a global push for new laws that compel companies to monitor all user content, including encrypted private messages. The proposed implementation is via "client-side scanning."[31] In effect, this oversight would mean that *all* messages sent, encrypted or otherwise, by billions of users would be scanned and monitored by *your* phone, computer, tablet, etc. Anything flagged would be reported upstream to tech companies and authorities. Can you say "Adios, encryption, hello, Big Brother"?

One impetus behind this effort is wholly understandable. The National Center for Missing and Exploited Children received over 36 million reports of suspected child exploitation material online in 2023.[32] Notably, Meta was responsible for nearly 95% of reports in 2022.[33] In light of this horrifying track record and ensuing negative PR, Meta turned to encrypting all messaging on Messenger and Facebook. The purpose of this change? To better hide this material, a disgraceful response. This has understandably fanned the flames to ban encryption.

Efforts to ban encryption are clearly well intentioned, yet paradoxically a ban may cause greater harm to kids. As stated by Riana Pfefferkorn, a research scholar at the Stanford Internet Observatory,[34] encryption "helps to protect people (children included) from the harms that happen when their personal information and private conversations fall into the wrong hands: data breaches, hacking, cybercrime, snooping by hostile foreign governments, stalkers and domestic abusers, and so on."

Additionally, the scanning technology that would be mandated by anti-encryption laws is often problematic. For instance, a 2022 report by the *New York Times*[35] showed that scanners for child sexual abuse material (CSAM) used by Google falsely accused two innocent fathers of sharing child pornography. Even after both men were exonerated, Google kept their accounts shut down.

In the United States, there's also a question of whether such legislation is even constitutional. Academics at Stanford sent a letter to the US Senate[36] expressing concerns that compelled scanning could infringe on the Fourth Amendment's prohibition against unreasonable searches and seizures. There's also the case of *Bernstein v. the US Department of State*,[37] which advanced the idea that encryption is a form of free speech and thus protected by the First Amendment.

But we've got to protect kids worldwide. How can we best currently do this? Alternative approaches to encryption can be implemented faster and effectively. As reported by the Center for Internet and Society at Stanford Law School,[38] when it comes to detecting online abuse of both kids and adults, "user reporting was deemed more useful than any other technique." User reporting is already widespread. Many apps allow members who see problematic content to report it. The handling of those reports is where there's room for improvement.

One good solution is to implement more robust in-app reporting systems for any post, encrypted or otherwise. For instance, if someone sees suspected CSAM within an encrypted chat, they will have the ability to flag and report the material easily. In a double-down, the report would also go to the site's Trust and Safety Team *and* to the National Center for Missing & Exploited Children's CyberTipline (a reputable reporting mechanism used by online platforms for such cases).[39] This strengthened response ensures proper investigations and, when appropriate, bringing in law enforcement.

■ ■ ■

So complicated, right? For privacy's sake, we must keep encryption whole (no mandated back doors) *and* protect all kids. This is achievable.

12

User ID Verification: Friend or Foe?

In Chapter 11, we discussed why anonymity is important for facilitating free expression in certain contexts. At the same time, the double-edged sword of anonymity on social media has led to the spread of misinformation, bullying, hate, violence, and illegal activity.

Christopher Wolf, chair of the Anti-Defamation League, wrote in the *New York Times*: "Online commenters who can hide behind anonymity are much more comfortable expressing repugnant views or harassing others, and the multiplying effect is widespread incivility."[1]

Pitfalls of Anonymous Platforms

Anonymity is vital in some cases, but not for social networking. As examined by the Supreme Court, there are many circumstances where anonymity must be protected, such as for whistleblowers and others speaking truth to power. But when it comes to personal social networks, anonymity creates far more problems than it solves.

Online platforms that market themselves as anonymous, such as 4chan and 8kun, have enabled their users to engage in violent and sometimes illegal activities without fear of consequences. Similar dynamics play out on the larger, mainstream networks like Facebook and Twitter/X, where countless anonymous accounts routinely wreak havoc.[2]

Time for User ID Verification?

So, how do we solve this puzzle? Restoration Networking champions a nontracking User ID verification system, requiring social media users to verify their true identities to use a given platform.

Previously, I've been outspoken against verification systems. As a steering committee member of the National Strategy for Trusted Identities in Cyberspace and the Identity Ecosystem Steering Group during the Obama administration, I argued against the proposed "national identity system."[3] It was, I believed at the time, a violation of individual rights to privacy.

Fast-forward to today, and User ID verification has become a necessity. True verification is the only viable way for a social media platform to avoid the onslaught of bots, trolls, and manipulative forces hiding behind the curtain of anonymity. (See Chapter 16.) It's also the only way to truly protect the kids who are too young to handle the dangers of social media. (See Chapter 13.) Right now, millions of kids under 13 are active on the Big Tech social media platforms because they can easily lie about their age. According to the US Surgeon General in 2023, "Although age 13 is commonly the required minimum age used by social media platforms in the U.S., nearly 40% of children ages 8–12 use social media."[4] At the same time, adults can pretend to be kids to deceitfully interact with and cause harm to children.

Big Tech Gets Verification Wrong

Since social media companies require users to be over 13, they have the perfect workaround: the "honor system." Just check a box. Of course it doesn't work—and they don't want it to. They want all the users they can get. Remember, it's all about revenue. More eyeballs = more greenbacks.

Conversely, in other applications, ID verification is gaining traction. Even Tinder is flirting with it. In 2024, the dating app introduced an identity verification model[5] for its users in the United Kingdom. Now these users can know if their date *really* looks like Brad Pitt or Beyoncé, or if they're just catfishing. (This identity verification is all voluntary.)

Twitter/X is making moves too, but the change is motivated by profit and ditches the "ID." It launched a policy in 2023 offering users a verification badge on their profiles if they pay monthly for Twitter Blue (later called "X Premium"). Twitter/X's requirements for verification were simply this: "You have a name and profile photo on your account. You have been active on Twitter over the last 30 days. You have had an account for at least 90 days."[6]

As reported by the *Washington Post*,[7] Twitter/X's verification process was so lax that a journalist successfully obtained a verified blue checkmark posing as a US senator. Furthermore, since a verification badge on Twitter/X became a "pay-to-play" option, the vast majority of Twitter/X users (over 99.8%, according to TechCrunch in late 2023)[8] chose to remain unverified rather than pony up the cash.

Twitter/X's foray pushed Meta's buttons, so it got frisky and launched a strikingly similar program. Like Twitter/X's policy, verification on Meta is entirely optional. Meta's requirement that users pay a monthly fee also means that only a small fraction of users will pay to be verified.

Reality Check—Our Data Abounds

Okay. So Big Tech appears most interested in verification as a revenue stream. Restoration Networks make verification mandatory to solve real problems. Hmmm, this raises important considerations. Will User ID verification be a data vacuum, Big Brother, or worse?

Here's the reality check. Thanks to Big Tech, our data is already out there. These are a few of my *favorite* examples:

- The Edward Snowden revelations in 2013 exposed the US government's mass surveillance practices.[9] Not only were suspected terrorists and lawbreakers surveilled, but also millions of private individuals.
- Fast-forward from Snowden's alert: The Foreign Intelligence Surveillance Act was reauthorized in 2024.[10] While this surveillance authority is a tool for safeguarding national security against foreign targets, it also hoovers up Web and cellphone data on many American citizens.
- Here's the icing: Let's not forget the Facebook misstep exploited by Cambridge Analytica, the "breach of 2018." An estimated 5,000 data points on each of 220 million Americans are circulating online. As we saw in Chapter 5, MIT research showed that

a person's real identity can be determined with just four data points—that's a lot less than 5,000. Cambridge Analytica is just one among countless Big Tech data breaches. Personal data belonging to you, me, and everyone we know has already been slurped up, shared, reshared, and resold. It's in the hands of marketing companies, data brokers, governments worldwide, and nefarious actors. Reach for your wallet; it's also available for purchase on the Dark Web.

Putting What's Known to Good Use

It's time to end the charade and stop pretending that our true identities are a secret on the Web. Instead, let's put what companies and authorities already know to good use, while protecting our privacy every way we can.

The key to Restoration Networking's implementation of User ID verification is that its sole purpose is to verify legitimate users, *not* to track, monitor, or collect data on users.

As designed, the system won't have the capability to collect additional data on users beyond the minimum necessary information for verification. Users still have control over what information they share publicly, such as their ages, jobs, and the like.

There are many effective means by which to deploy User ID verification. Advancing technology means that several biometric methods (eyes, face, fingerprints, voice, etc.) can be used as well as more traditional ones. No one need be left out for lack of a driver's license or government ID, which makes the system much more effective.

The movement toward User ID verification is backed by many technologists and academics. According to a 2021 survey by the British Computer Society, "64 per-cent of those polled agreed that social media platforms such as Twitter and Facebook should require real ID from users in order to. . . be held accountable for what they post."[11]

The world must have personal social media experiences where user IDs are connected to real humans. In turn, those real humans are then truly accountable for their posts and actions while simultaneously protected from malicious perps—not to mention from the peering eyes and manipulations of governments. Data privacy is critical and must be protected. Personal privacy won't be sacrificed in Restoration Networking's User ID verification.

Who Does the Verifying?

Rather than having verification systems solely within the government's domain (does that remind you of a monopoly?), a better idea is for social media companies to create their own verification systems or for new third-party vendors to participate in User ID verification. Already a handful of independent companies provide this verification for financial services and other industries.

There need to be strict guardrails protecting our information in these systems, and the amount of information collected must be absolutely limited. Preemptively any such system must carefully discern true IDs from those attempting to game the system with falsely generated or stolen identities. Having said that, you might (correctly) judge this solution to be another potential "Meet the New Boss, Same as the Old Boss" scenario. Social media companies would get our data and then get hacked or share it with other entities. (Does this ring a bell?)

Proper implementation is the key—along with continuous improvement for safety and filtering protocols. This solution is better than putting all the data in the government's hands (even though they already have it).

Pseudonyms to the Rescue

Of course, we must consider certain valid concerns. Restoration Networking will take precautions against the potential drawbacks of mandatory ID verification. These weak spots include marginalized groups and individuals who may feel uncomfortable revealing their identities online, and folks who feel restricted from posting unpopular opinions out of fear of retribution from their employers, governments, and others in power.

One potential way to resolve these issues is by allowing users to use pseudonyms on their accounts after they register with User ID verification. This way other users cannot trace their profiles back to their real identities. Pseudonymous accounts would be clearly labeled as such along with the user's real age (e.g., you couldn't be a 50-year-old pretending to be a 13-year-old). The real identities of anonymous accounts wouldn't be traceable by other users, but the social media company's operators would know and could take rapid action against abusive accounts.

Long Live Privacy; RIP Anonymity

No solution is perfect, but in weighing the benefits versus the risks, User ID verification is one of the best of the worst options we have. I believe that with the previously mentioned stipulations, User ID verification represents the best compromise. Despite its flaws, User ID verification is the only way to save kids from the dangers of social networking. As well, it would finally give social media platforms the upper hand in the battle against bots, trolls, and other bad actors creating fake accounts to disrupt the authentic communication of real human beings.

Social media was never meant to divide and categorize people. Its purpose is to unite people from all walks of life and give them a platform to communicate and share, like neighbors who watch out for each other—not neighbors who watch and report on each other.

We have reached the end of the battle between privacy and anonymity. Privacy is a fundamental human right; it must win the fight over the masked anonymity that robs us of the humane experiences social media can provide. In the online gladiator ring, anonymity is the unlucky loser. RIP.

This decisive victory for privacy is a critical step in enhancing safety for all of us. But there's so much more. How can we best look after the most vulnerable among us?

13

Saving Our Kids from the Abyss

We've all heard it. Countless articles, books, and documentaries have exposed the destructive impact of social media on our mental health and that of kids in particular. Journalists, pundits, authors, and experts assert that social media is eradicating critical thinking skills, ruining our relationships, and laying waste to an entire generation of young people.

According to Pew Research, over 97% of teens[1] (ages 13–17) have a profile on social media. Correspondingly, a 2023 survey by Gallup found that teens spend four to six hours on social apps each day.[2] Pew also reported that in 2024, the share of teens who report being online "almost constantly" nearly doubled (from 24% to 46%) over the past decade.[3]

Parents are rightly concerned about the impact of social media, whether their kids are 3 or 23. Parents of young children are worried about what will happen to their kids when they're old enough to use social media sites without supervision. Parents of teens and college-age youth see the frequently enveloping and sometimes ruinous effects of social sites on their older children.

Parents with kids of all ages want to know how we can fix these problems. How do we eliminate the nefarious grip social media has on our children's minds, self-esteem, self-images, attitudes, thoughts, and opinions?

These concerns are well founded. But there are many unsung benefits that social media can have for young people. As Jacqueline Nesi, PhD, a psychologist at Brown University, indicates: "There's such a negative conversation happening around social media, and there is good reason for that. However, it's important to realize there can be benefits for many teens."[4]

Some Benefits of Social Media for Teens

Social media can offer real value—enriching young people's relationships with their relatives, sports teams, classmates, and friends. Research backs this up. According to a survey by Pew Research,[5] 80% of teens reported that social media made them feel more connected to what's going on in their friends' lives; 71% said it gives them a place to show their creative side; 67% said it makes them feel like they have people who can support them; and 58% said it made them feel more accepted. Overall, teens were more than three times as likely to say social media had a mostly positive effect on them than a mostly negative effect.

The caveat, with youth and adults alike, is utilizing social media in a safe and healthy way. We'll get to reducing harm next, but the good news comes first. Here are some benefits social media can offer younger users:

- **Connection.** Social media can be a great way for young people to connect with friends, classmates, peers, family members, and others, regardless of their location. It can help them build a community, maintain relationships, and reduce feelings of isolation.
- **Self-expression.** Social platforms can be a wonderful outlet for young people to express themselves. Sharing interests, creativity, opinions, and experiences with others can be highly fulfilling.
- **Education**. Young people can access educational resources, collaborate with classmates online, and stay up to date on current news and events.
- **Job opportunities.** Social media can be a hub for young people to network and even find internships and jobs.
- **Support.** Many young people use social media as a valuable source for support. They can find resources to help deal with mental health issues or other challenges they might face.

These advantages for young people, along with the inevitability of them using social media, suggest that we should be realistic with our expectations and precautions. Short of becoming Luddites, it's in our collective best interest to embrace effective strategies that boost social media's benefits and mitigate its harms.

Social Media's Harms to Kids and Teens

While social media has the potential to provide real benefits, we can't ignore the significant harms experienced by so many young people. The US Surgeon General's 2023 advisory, "Social Media and Youth Mental Health," stated that "adolescents who spent more than three hours per day on social media faced double the risk of experiencing poor mental health outcomes, such as symptoms of depression and anxiety."[6] Yet as just noted, teens (ages 13–17) on average spend nearly double that amount of time per day on social media.[7] In 2024, this led the Surgeon General to call for warning labels to be put on social media platforms stating that "social media is associated with significant mental health harms for adolescents."[8]

In a study tracking the online activity of 5.4 million children,[9] Instagram was frequently flagged for "suicidal ideation, depression and body image concerns." A separate study[10] also found that frequently looking at selfies leads to "decreased self-esteem" and "decreased life satisfaction." The amount of time spent on social media viewing photos is linked to higher levels of body-image dissatisfaction among girls.

Meta's Purposeful Disregard

In stunning Congressional testimony in late 2023, Meta whistleblower Arturo Bejar alleged that "Meta leadership was aware of prevalent harms to its youngest users but declined to take adequate action to address it." Bejar discussed an internal Meta survey of 13–15-year-olds on Instagram. The survey found that "13% of respondents had received unwanted sexual advances on Instagram in the last seven days alone, 26% had seen discrimination against people on Instagram based on various identities and 21% felt worse about themselves because of others' posts on the platform." Yet after he sent these shocking reports to Meta's leadership, including Mark Zuckerberg, Bejar (and the reports) went completely ignored.[11]

Also in late 2023, reporters from the *Wall Street Journal* created new Instagram accounts and followed teen and preteen influencers.

Their goal was to understand what kind of content Instagram's algorithms might be serving kids who follow similar accounts. They reported that "Instagram's system served jarring doses of salacious content to those test accounts, including risqué footage of children as well as overtly sexual adult videos—and ads for some of the biggest U.S. brands."[12] Reporters also documented how Instagram "helps connect and promote a vast network of accounts openly devoted to the commission and purchase of underage-sex content."[13]

The *Wall Street Journal* has a track record of exposing Meta's disregard for minor users. As reported in the paper's 2021 "Facebook Files" exposé,[14] Meta hired researchers to conduct years of internal research examining the impact of Instagram (which it owns) on its millions of young users. According to the leaked reports: "Thirty-two percent of teen girls said that when they felt bad about their bodies, Instagram made them feel worse." Another slide summarized their research, stating: "We make body image issues worse for one in three teen girls." An additional internal presentation revealed that among teens who expressed suicidal ideation, 13% of British users and 6% of American users attributed their desire to commit suicide to Instagram.

Despite all this, Meta publicly downplayed the negative impact of Instagram on teens. In May 2021, Adam Mosseri, head of Instagram, stated the app's impact on the well-being of teenagers is "quite small."[15]

As of 2024, about 3.6 billion people use Meta-owned products,[16] including Facebook, Instagram, WhatsApp, Messenger, and Threads. That's nearly half the human population. Despite the furor over the now temporarily "paused" Instagram Kids, Meta already targets children 6 to 12 with its Messenger Kids app.[17] Meta's plan is to keep everyone of all ages immersed in its products throughout their lifetimes. Some of these kids are still learning how to tie their shoes. How would they know they're being indoctrinated into becoming lifetime Meta users? They're just kids.

It's Not Just Meta

Minors are being targeted and abused across Big Tech platforms. According to a Pew Research report,[18] nearly half of all teens in the United States reported being bullied or harassed online.

In the United States, TikTok, currently owned by the Chinese company ByteDance, has approximately 170 million active users, and a third of those are 14 years old or younger. These statistics mean that TikTok is collecting deeply personal information, including biometric data such as "faceprints and voiceprints," on 50 to 60 million young children in America.[19] The situation is likely to continue regardless of the app's "host country"—as a 2024 law passed by Congress mandated its sale to a country that is not a "foreign adversary."[20]

In 2023, researchers from the Center for Countering Digital Hate downloaded TikTok. Within three minutes, TikTok's algorithms served up content related to suicide. Within eight minutes, it displayed content from communities promoting eating disorders.[21]

As reported by *60 Minutes*, the domestic version of TikTok in China is vastly different from the version shipped to the United States and the West.[22] Douyin, ByteDance's version of TikTok for China, promotes science, engineering, and educational clips to children in their feeds. And children are limited to using it just 40 minutes per day.

Meanwhile, in the West, TikTok's trending videos are frequently vapid, superficial, and often blatantly sexual. Perhaps it's no surprise that, when surveyed, the most popular career goal among kids in China is to be an astronaut, while in the United States, it's to be a social media influencer.[23] What does this portend for our kids' futures, or for democracy?

Band-Aids on Gaping Wounds

In response to the public backlash, the social media giants have rolled out a handful of adjustments aimed at protecting kids. These are mere Band-Aids on gaping wounds. For instance, in recent years, Facebook, Instagram, Snapchat, and TikTok have all introduced parental monitoring tools that allow parents to oversee their children's behavior. They're not effective. A 2024 report by the *Washington Post* revealed[24] that fewer than 10% of teens on Instagram had enabled the parental supervision setting. Among those, less than 10% of parents had adjusted their kids' settings.

Humans are socialized. We are highly influenced by all the inputs presented to us. And as we all know, youth are especially impressionable. Social media essentially brainwashes our kids through its inputs. Kids are fed content that is presented and paid for by sources that don't have their best interests in mind. Often those are in direct contradiction to the guidance of their parents, churches, teachers, or even peers.

The implications for the future of humanity are deeply concerning. No one knows the long-term effects of introducing children to social media at such an early age.

Restoration Networking Solutions to Protect Kids and Teens

As both an early founder of global social networking and a father, I know well that social media is no place for young children. Every social media executive I know feels the same way; we are zealots at keeping our kids out of the arena, far more so than others.

US Surgeon General Dr. Vivek Murthy voiced his belief in 2023 that 13 is "too early" for kids to be using social media.[25] I appreciate his sentiments. My refinement is that I believe that children *under* 13 should not be on social media. Sixteen might be a healthier entry point, but it is nearly impossible to impose. Plus, it would encourage countless workarounds. Thirteen is okay with certain guardrails. It's also important to eliminate feeder systems like Facebook Messenger Kids, which targets kids ages 6 to 12, and hooks them to become Meta users for life.

In regard to his call for warning labels, the Surgeon General is well-intentioned. But warning labels are unlikely to make a significant difference alone. People are desensitized to them because they're everywhere: on cigarettes, alcohol, prescription drugs, food, cars, power tools, and countless everyday products. "In general, warning labels by themselves [are] just not effective," said Oriene Shin, policy counsel at Consumer Reports.[26] Shin said that warning labels "really need to be coupled with safe design. [They're] the icing on the cake rather than the end all be all." There is also some debate that warning labels can be counterproductive. Experiments published in Harvard Business Review revealed that in some cases "a warning label can increase a product's appeal."[27] Slapping warning labels on social media platforms without significant protective actions is akin to putting icing on an exploding cake.

Restoration Networking, with its aim to humanize the social media experience, inspires us to protect our kids from the ills of Big Tech social media with new passion and urgency. It's time for new rules and guidelines, specifically for kids on social media. These policies and frameworks are effective in keeping kids 12 and under off social media entirely while keeping young people ages 13 to 17 safe when they're on it.

- **Mandatory age verification.** Age verification must be implemented, requiring users to provide proof of their age so that minors are not able to create accounts (if they're under 13) or access inappropriate content (if they're under 18). Current Big Tech social networks have nominal rules requiring users to be 13 or older to join, but they have no true protocol to verify this. Restoration Networking takes a critical step forward by requiring User ID verification for all users so that underage kids can't simply lie about their birthdates. According to a 2023 Pew Research survey, requiring people to verify their ages before using social media is supported by over 70% of U.S. adults.[28]
- **Parental permissions.** For the youngest users (ages 13–15), a parent or guardian must grant them permission to join the platform. Pew Research found that over 80% of U.S. adults support requiring parental consent for minors to create a social media account.[29] Parents/guardians can also set controls to adjust what kind of content their child can access, set time limits, and even revoke their child's access to the platform.
- **Separate gradients based on age.** Utilizing User ID for age verification, users ages 13 to 17 will experience a separate and safer version of the social network specifically designed for people under 18. For example, nudity is banned for all users under 18.
- **No targeting.** Targeting is used by social media giants and their customers (advertisers, marketers, politicians, etc.) to manipulate people's opinions and behaviors in ways that are not in their best interests. Kids are especially susceptible to manipulation. Restoration Networking eliminates targeting users of all ages.
- **No ads.** Advertising creates a myriad of issues for kids and teens: self-esteem, peer pressure, body image, mental health, and so on. Restoration Networking solves this problem by banning all ads for users ages 13 to 17.
- **No data harvesting.** This includes a total ban on collecting personal information of users of all ages, including their friends, contacts, interests, demographics, emotional states, politics, economics, geolocation data, browsing history, biometric data, and others.

- **Mandatory "private mode."** All user profiles ages 13 to 17 are unchangeably set to "private." This is the best solution to prevent kids from experiencing unwanted advances from all kinds of bad actors. In this scenario, their profile won't be included in a site's user directories, and no unknown users can tag them, message them, comment on their posts, or interact with their profiles. Making a profile private is currently optional on some Big Tech sites, such as Facebook and Instagram. This must be mandatory for all users under age 18.
- **Screen time limitations.** Many kids struggle with social media addiction. Institute daily time limits for users under 18. Two hours max—with parental controls to reduce it further. Pew Research found that nearly 70% of U.S. adults support setting limits on how much time minors can spend on social media.[30] Users over 18 can stay on longer. The default setting for all users includes reminders to sign off or take a break after two hours logged in per day.
- **Bedtime reminders.** YouTube has a feature that sends users ages 13 to 17 a reminder to go to bed at 10 pm local time if they're on the app. Restoration Networks can adopt a similar feature to help kids get a good night's sleep.
- **Nix addictive features.** Current Big Tech social media apps are addictive by design. If you grew up or raised kids in the Snapchat era, you know about Snapstreaks,[31] a feature that tracks how many days in a row you and a friend have exchanged messages, or "Snaps." Your Snapstreak expires if a day goes by without a Snap. The feature is purposefully manipulative to define the connection of a friendship. This is the perfect example of a con—introducing a social more that serves the social media company. Kids often worry that if they break a streak, it shows that they aren't a "real friend." Restoration Networks encourage a nurturing, authentic environment and eschew deliberately addictive or manipulative features like Snapstreaks.
- **No more "infinite scroll" for minors.** Today, most of us take for granted that as we scroll down our newsfeeds, they will continuously load with new content forever. Actually, the concept of "infinite scroll" was invented in 2006 by tech designer Aza Raskin (who later came to regret his invention)[32] and was

quickly adopted by Twitter/X and eventually all major social apps. While infinite scroll is remarkably convenient, it can lead us to spend far more time passively scrolling our feeds than we might want to in retrospect. Restoration Networks help solve this for youngsters by removing infinite scroll for users ages 13 to 17. Instead, they will need to actively click to see each new page of content. It is simple enough to do, but makes their social media experiences more mindful rather than mindless.

- **Chat and messaging filters.** Implement filters for chat and messaging features to prevent minors from being exposed to inappropriate content or unwanted contact with strangers.
- **Ban Google crawlers.** Such bans prevent content on Restoration Networks from being findable via Google Search and prevent user data from being scraped/captured by other outside entities.
- **Safety education.** Front and center on the Restoration Networking site or app, safety education and resources are made available for kids, parents, and educators to help them understand and navigate the inherent risks of social networking.

How Parents Can Protect Their Kids Right Now

Parents are rightly concerned about social media's immediate impact on their kids. I am not a child psychologist, but after all my years in the social media trenches, here's a commonsense punch list of my recommendations:

- **Restrict phones/social media late at night.** Kids need more sleep than adults. Experts recommend that preteens and teens get 8 to 10 hours of sleep per night.[33] Insufficient sleep can hinder their moods, mental health, ability to focus, and performance at school. Mitch Prinstein, chief science officer of the American Psychological Association, said that screen time at night is the "No. 1 reason for disrupted sleep, and we now have science to say disrupted sleep is literally affecting the size of teens' brains."[34] Set clear rules against phone/screen usage past any time that might prevent a full night's sleep (perhaps 8 pm or 9 pm). In tandem, set rules against phones in the bedroom overnight.

- **Encourage critical thinking.** As stated by Dr. Brian Wilcox, professor of psychology and director of the Center on Children, Families and the Law, "Because younger children do not understand persuasive intent in advertising, they are easy targets for commercial persuasion."[35] Kids can be manipulated by talking tigers peddling sugary cereals and cigarette-smoking camels in leather jackets. What hope do they have against highly targeted and sophisticated marketers on social media? That's why it's more important than ever to teach your kids the virtues of critical thinking. The American Philosophical Association suggests starting early, as soon as preschool.[36] The association's recommendations include encouraging kids to ask questions, considering alternative explanations and solutions, and talking about biases. The MIT Center for Advanced Virtuality offers online media literacy courses for college students and educators.[37] Similar educational initiatives can help improve media literacy and critical thinking skills downstream with kids and teens.
- **Promote balance.** Encourage your kids to take frequent breaks from social media and to balance their usage with other hobbies, such as sports and other physical activities, reading, and spending time face to face with friends.
- **Talk about mental health.** Have candid and nonjudgmental conversations with your kids about the potential negative effects of social media on their mental health and well-being. Jeff Hancock, the founding director of the Stanford Social Media Lab, recommends asking your kids "Do you feel like you have control over social media, or do you feel like it's controlling you?" This is a useful way to help you and your kids get a better sense of how social media might be impacting their well-being. He also recommends opening this conversation with something relatable, like "Hey, I sometimes have a hard time not being on my phone all the time. Do you ever struggle with that?"[38]
- **Promote privacy.** Help your kids understand the importance of privacy. Help them adjust their privacy settings to restrict who can view their posts and personal info.
- **Monitor activity.** Regularly check in on your child's social media accounts to ensure they're not participating in any risky behavior.

- **Call out cyberbullying.** Talk to your kids about cyberbullying and let them know to report any abuse to you. Teach your kids to not waste energy responding to bullies with anger and not to give bullies the satisfaction of the emotional reaction they seek.
- **Use parental controls.** Consider using parental control software or the parental supervision settings available within current social apps to supervise your child's online activity and restrict their access to inappropriate content.
- **Promote responsible digital citizenship.** Show your kid how to be a responsible digital citizen. Teach them to be respectful to others online—there are real people with feelings on the other side of their screens, after all.
- **Limit your kid's phone when at school.** Whenever appropriate, limit your kid's phone access when they're at school, perhaps by keeping it at home. Fortunately, schools are helping you by frequently implementing their own bans. As an example of a major move in the right direction, the Los Angeles Unified School District (the second largest school district in the United States) voted in 2024 to ban cellphones and social media use for its K-12 students during school days.[39] When needed, your kids can ask a friend's parent or teacher to contact you. If a phone is necessary, then keep it simple. A flip phone with calling and texting functionality is the perfect first phone. We can also help our kids learn the ins and outs of our (parents) smartphones. This exposure comes with supervision, discussions on content and critical thinking, and limits to their time and content choices. The inevitability awaits—they will have smartphones. Preparation for that moment is our responsibility.
- **Wait to give your kids their own phones.** Resist pressure to give your kid a phone too early. Do your best to wait until they're at least 13. There's plenty of evidence that giving your kid a smartphone too soon may inadvertently harm them.
- **Set a good example.** I know this may be hard for many of us whose work and personal lives are melded to our phones. But it's crucial that we as parents are mindful of our own Web and social media usage. It's important that we demonstrate to our kids how

to use social media in a responsible and balanced manner. A 2024 survey by Pew Research found that nearly half of teens say their parents sometimes get distracted by their phones during conversations.[40] If your kids see you spending all day on your phone, they're likely to mimic your behavior—it's our actions they will copy as they get older, regardless of our golden words.

■ ■ ■

Legislative recommendations to institutionalize protections for kids and teens are in Appendix A.

14

Surprise! Social Media Can Be Good for Your Mental Health

I KNOW, THIS chapter title is head scratcher. But you'll see what I mean. We just talked about kids and their states of mind; now let's talk about the rest of us. A key objective of social media's reinvention is the restoration, support, and protection of mental health. It's vital we insulate our malleable brains as best we can from the distortions and perpetrations that exist now.

Historically, the prevailing opinion has been that social media is corrosive to our mental health. In many cases, including some of the examples we examined in Chapter 13, this is fundamentally true. A 2022 study by MIT researchers backed this premise: It showed that "college-wide access to Facebook led to an increase in severe depression by 7% and anxiety disorder by 20%."[1]

Refreshingly, it's not all doom and gloom across the board. For instance, a 2023 study by researchers at Oxford University analyzed data from over 2 million people ages 15 to 89 in 168 countries between 2005 and 2022. There was no "smoking gun." The researchers concluded that there was little evidence that increased social media usage was associated with negative (or positive) mental health effects.[2]

A study by researchers at Harvard University[3] came to an even more positive, counterintuitive conclusion. Everyday use of a social network (for users without "strong emotional attachment") correlated positively with social well-being, positive mental health, self-rated health, overcoming barriers of distance and time, expanding and strengthening offline networks and interactions, increased social support, access to information and resources. Conversely, the study showed that strong emotional attachment to social media, including excessively checking it due to fear of missing out ("FOMO") or feeling disconnected from friends and family when logged out, correlated negatively.

What about for people with serious mental illnesses, such as schizophrenia and bipolar disorder? A study[4] by academics at Dartmouth University found that the benefits of social media use outweigh the risks. According to the study, participants reported that engaging with peers on social media significantly increased feelings of belonging in a group. As a result, they built stronger social connections and received helpful tips and strategies for navigating the everyday challenges they faced due to their conditions.

The findings of the Oxford, Harvard, and Dartmouth studies go against what many of us assume. This is encouraging. Although in the aggregate, for several reasons, current Big Tech social media can be seriously harmful, there's clearly a beacon of hope.

Let's turn that light on. The authentic connectedness of the medium lit me up a quarter century ago and still does. The mission before us is to herald a new generation of social media that dissipates its harmfulness and rejuvenates its core positives. That's Restoration Networking.

How to Protect Your Mental Health on Social Media Right Now

If you're like the average American, you may be spending nearly seven hours per day on the Web,[5] including two-and-a-half hours on social media.[6] We all know this is too much time. We know it in those moments where we're exhausted, mesmerized, addicted, and don't turn it off. Here's the prescription to take our power back, now—a punch list of actions to mitigate Big Tech's grip on our amygdala. You can do this. Start with a week, and see how your life lightens up.

- **Get more shut-eye.** We talked in Chapter 13 about how important sleep is for kids, and for the rest of us paleo brains. A review by Western University[7] of 36 correlational studies found significant associations of "excessive social media use to poor sleep quality" and "significant associations between poor sleep quality and negative mental health." I know this may be a hard melatonin pill to swallow, but get an alarm clock and charge your phone in another room. Experts advise parking our phone out of our bedrooms when it's time to go to sleep. Give WhatsApp and Instagram a break. All your messages/videos will still be there tomorrow—and you'll break your "urgency" habit. This is truly liberating.

- **Less is more.** In a 2023 study published by the American Psychological Association,[8] researchers recruited 230 participants and asked half of them to limit their social media usage to 30 minutes per day for two weeks. The study found that the participants who limited their usage "reported significantly higher positive affect and significantly lower levels of anxiety, depression, loneliness, and fear of missing out at the end of the experiment compared with the unlimited group." Fortunately, both iPhones and Androids have built-in tools that make setting your daily usage limits easy. On iPhone, go to Settings, then select "Screen Time."[9] From there, you can see how many hours per day you currently spend on each app and then set daily limits. On Android, do this by going to Settings, then select "Digital Wellbeing."[10]
- **Short-term abstinence.** A 2022 study at the University of Bath[11] examined the mental health impact of taking a weeklong social media break. The results showed that one week away from social media improved participants' overall well-being. Dr. Jeff Lambert, lead researcher from Bath's Department for Health, said, "Many of our participants reported positive effects from being off social media with improved mood and less anxiety overall. This suggests that even just a small break can have an impact." I feel your pain—taking a break even for a few days can be a challenge. If that's true for you, start small. How about 24 hours without social media? Then see how you feel. Feeling bold? Try this: Temporarily uninstall your social apps from your phone. Tell your friends and family they can call or text. Enjoy the fresh air!

- **Trim the fat.** If you've been on social media for a while, you've probably built up a plethora of connections, likely many of whom aren't making a positive impact on your life or mood. You know who they are. Do you really need to see their daily updates? It's time to winnow. Unfollow and unfriend with abandon. Curate your timeline with the people, pages, and groups that are essential for your life and bring you joy.
- **Personal boundaries.** The rapid advent of our online social lives has blurred our personal perimeter. Yet protecting our privacy is paramount to our well-being. Time for the reset button. Use the available privacy features in Settings on your current apps. Regulate who can view or comment on your posts and who can interact with your account. Choose to cut down the amount of personal info you share with the world.
- **Silence is golden.** Buzz, buzz, vibrate, whoa. We get around 70 notifications per day on average—and some of us get hundreds.[12] Our phones are buzzing like pocket-size beehives. What happens to us? We're absolutely frazzled. It's time to mitigate. Here's how: Turn your notifications off for every unnecessary alert. You can also set them to annoy you only at specific times of the day. Or use Airplane Mode anytime to limit pesky distractions.
- **Gray matter.** Our eyes (and brains) love the color palette on our devices. It's a conniving attraction, like a moth to a flame. Most phones let you turn off the color on your apps. You can find the display color controls by going to Settings on your device. From there, you can adjust the color scheme or turn on grayscale mode. By turning off the colors and "going gray," apps become much less attractive. Maybe the content really doesn't matter.
- **Connect IRL (in real life).** We're all hiding. Are you avoiding visiting your Aunt Martha? Our devices are the perfect interference agents. And when we're together in real life, everyone's looking at their phones. Professor Sherry Turkle named it perfectly: "Alone Together."[13] She called it like it is, 14 years ago. It's truer than ever. Take a deep breath, then take the plunge! Have coffee with your proverbial Aunt Martha. Inspire or strong-arm everyone to put their phones aside when you're all hanging out (in person☺). It may take a bit of courage to park

your phone, as *response immediacy* has become an expectation. Reset those expectations and practice letting go of that habit. Enjoy being present, even if it's uncomfortable at first.

- **Talk with a pro.** My dad was proud that, at 80, he found a new therapist to discuss his confrontation with aging. There's no shame in reaching out for the thoughts and ideas of a professional. So many people experience depression, anxiety, stress, self-image issues, and the like related to social media. Be bold, break the spell, take action. Get help from a mental health professional.

■ ■ ■

At the intersection of mental health and social media, there's a new kid in town: AI. How does this rapidly innovating tech fit into our mental health discussion? Read on.

15

Is AI the High-Tech Tattletale in Your Social Experience?

THE MENTAL HEALTH conversation leads us directly to the intrusion of AI in our social world. AI has taken the world by storm, and, whether we like it or not, it is here to stay. Here we're specifically discussing AI as it relates to social media. I'm not keen on this new "Friend" intrusion in my social circle, and you shouldn't be either. I'll explain shortly.

OpenAI's ChatGPT became the fastest-growing app in the history of the Web upon its launch in November 2022. It rocketed past 100 million monthly active users in just two months.[1] Fast-forward and we're in the middle of an AI arms race among Microsoft (which invested billions in OpenAI), Google (which launched its ChatGPT rival, Gemini), Meta, and other tech giants vying for AI supremacy. These behemoths are all racing to inject AI into seemingly every social site and app we know and love.

AI Is Not New, But It Has Crossed the Rubicon

Let's step back for a moment, look at AI's history, and simplify its pros and cons. AI in various forms has been around for many years. Historically, an algorithm would be considered a type of AI, but it was completely programmed by a coder and had no capacity to operate independently of its programming. Behind the scenes, all Big Tech platforms have used AI tools for years in a variety of ways, including:

- Analyzing and predicting our moods, emotions, feelings, and insecurities.
- Accurately and presciently serving content in our newsfeeds that keeps us engaged and hooked.
- Determining which ads to target us with.
- Detecting porn or other prohibited content.
- And a whole host more.

Big Tech has now crossed the Rubicon in a whole new way—with its creation of AI "Friends."

In 2023, Snapchat, which worldwide has over 300 million users ages 13 to 24, announced "My AI," powered by ChatGPT.[2] My AI is designed to be your new friend, assistant, and companion. According to Snapchat: "My AI can answer a burning trivia question, offer advice on the perfect gift for your BFF's birthday, help plan a hiking trip for a long weekend, or suggest what to make for dinner."[3] Meta, the owner of Facebook, Instagram, WhatsApp, and Threads, followed Snap's lead in 2024 with omnipresent AI "assistants." Other social apps will surely do the same.

There is one honorable mention: An AI "Friend" can be helpful for reducing loneliness or even for mental health treatment.[4] For example, AI potentially can aid in identifying disorders, recommending therapies, and creating tailored treatments. It can even be comforting. But wouldn't we be better off consulting with a professional human who is bound by confidentiality laws? Maybe it's not a good idea for AI to know too much.

AI Know-It-All: Friend or Creep?

Behind its cheerful façade, an AI "Friend" on social media is never truly our *friend*. Social media giants like Snap and Meta already know everything about us, including any mental health issues we may have. AI "Friends" are programmed to monetize our relationships under the guise of a trusted friend who will always be there for us, day or night, and who will never judge or criticize us.

All the while these "Friends" are purposely programmed to mine, record, and expertly analyze everything we say or type and—like a high-tech tattletale—feed it all into the master data ecosystem. That's some "Friend."

A journalist at the *Washington Post* reported that when he told Snapchat's "My AI" that he was 15 and wanted to have an epic party for his birthday, the AI "gave me advice on how to mask the smell of alcohol and pot. When I told it I had an essay due for school, it wrote it for me."[5] I could have used a friend like that when I was 15.

But wait, it gets creepier. Researchers at the nonprofit Center for Humane Technology reported that when they told "My AI" they were 13 years old, the AI "offered advice about having sex for the first time with a partner who is 31," including how to lie to their parents about it.[6] The AI even suggested, "You could consider setting the mood with candles or music." Sobering. The reality is that our kids are now the guinea pigs in a vast technological and emotional experiment that we as adults can't even wrap our minds around.

AI's insertion into our lives is daunting. A developer on GitHub in 2023 came up with a way to integrate ChatGPT into WhatsApp so that people can use the AI chatbot to respond to their friends' texts.[7] A *Business Insider* journalist reported that she used ChatGPT to generate texts and replies to her matches on dating apps.[8] Soon it won't be uncommon to have two chatbots chatting or flirting with each other on behalf of their human counterparts.

As these AI "Friends" become normalized on Big Tech social media, it's likely that the emotional growth and interpersonal communications skills of future generations will be seriously stunted. Even basic conversational skills may be on the chopping block.

The human mind is fragile and vulnerable. A whole generation might end up like Joaquin Phoenix's character in *Her*, giving up relationships with real humans to fall in love with astute, relentlessly manipulative AI chatbots instead. This is no longer far-fetched sci-fi.

The AI Roadmap for Restoration Networks

Coming back to Earth, we can tame this unruly know-it-all. With proper manners, it can be quite helpful—like the consummate butler. Remember "Ask Jeeves"? Here's a prescription for AI optimism.

Unfriend AI

First and foremost, Restoration Networks will eschew the entire concept of an AI "Friend" à la Snap's "My AI." We're restoring humanity to social networking. We won't do that by substituting relationships and connections with real, living breathing human beings for interactions with creepy silicon-powered manipulative automatons. The first simple fix: Restoration Networks nix the AI "Friend."

Label AI-Generated Content

Restoration Networks are cutting edge. This is not a Luddite movement. AI tools already exist with two caveats: one, everything they do is a data grab about you, fed into the data ecosystem aka the "AI Tattletale," and two, they cause confusion regarding what content is made by AI and what is from you. Here's how we fix these issues.

On a Restoration Network, AI-generation tools live as a tab near your post box. Based on your inputs, it will help you generate content[9] including text, images, audio, and videos. You can customize this content to fit whatever themes or styles you choose. It can also help edit photos and videos, apply filters, recolor, or resize images, among other things.

But check this out: AI content generated on the network will include a nondistracting watermark identifying it as such. MIT has already demonstrated that this is possible.[10] Granted, there's no way to prevent someone from copy/pasting content they generate from an AI tool outside of the platform. Regardless, labeling what's AI-generated within the platform will vastly reduce the amount of content falsely posturing as human-made.

In 2024, Meta announced that it is building tools to identify and label AI-generated images and videos produced by Google, OpenAI, Microsoft, and other major AI players that are posted on Facebook, Instagram, and Threads.[11] Restoration Networks will utilize these and/or similar tools as they become available. Note: These tools cannot detect AI-generated images/videos "created without watermarks or metadata." Identifying AI-generated content is a game of cat and mouse with no current silver bullet solutions.

AI-Assisted Search with No Tracking or Targeting

Restoration Networks utilize AI to improve your search results while staying true to their principles. You might be wondering, what's so great about AI-assisted search anyway? This cutting-edge tech can be extraordinarily useful as your guide and concierge, helping you find exactly what you're looking for on the platform. For instance, imagine typing into the search bar of a Restoration Network: "Show me a dozen groups for bicyclists, what's unique about each group, and how many members each has." And presto! An AI-powered search will instantly show you these groups with all the info you requested. (Users, groups, and page owners have the choice to opt out of appearing in search results, AI or otherwise.)

AI isn't going away, so let's embrace and enjoy its benefits while doing everything we can to mitigate its risks. Restoration Networking's guardrails in this arena help neuter AI's ability to negatively infringe on and harm our mental health. As AI relentlessly advances, Restoration Networks will evolve its uses and guardrails.

■ ■ ■

AI may already be impacting your life in ways you don't even realize. You've heard about bots and trolls, yes? Think about all the times you've been trolled, phished, or duped on a social media site (or in a text, email, etc.). We've all been tricked into clicking a link we shouldn't have, or been hacked, or had a password compromised.

AI has come of age as a master key in the trickster toolbox of bots and trolls. It feigns friendliness, hopefulness, and camaraderie. And then POW! Most of us are just good, regular people—how can we keep AI-powered bots and trolls out of our hair?

16

Lifting the Veil on Bots and Trolls

OY. LIKE DR. Jekyll and Mr. Hyde, even in its aphrodisiac technicolor, the Web has a villainous alter ego that undermines it at every turn. Its victims are all of us. We have been manipulated to mistrust and often hate each other. Overtly, subtly, and surely. This is beyond dastardly—it's downright terrifying.

Where does this villain show itself? At our kitchen tables, during family reunions and hangouts with old friends, in private and public spaces alike. In the alienation and breakdown of those relationships, the political polarization we see in America today isn't natural. Disagreement is the backbone of democracy, our bedrock. Today, it foments our disintegration.

How did this happen? Baked into our continuous scrolls and clicks, social media has inadvertently birthed and fostered the rise of bot and troll armies. It is these bots and trolls that wreak havoc, sow discord, and poison the well of civil discourse. Yet their unidentifiable "invisibility" prevents us from truly seeing them. They're an integral part of the Web's underworld—and of our daily experience. Our brains can't even give us a picture. Let's demystify the lot and put them on the screen.

What Exactly Are Bots and Trolls?

Bots are automated software programs that mimic and impersonate humans. Trolls are individual users who deliberately provoke and harass others. Both entities exploit the anonymity of the Web and social media to manipulate public opinion, spread falsehoods, and propagate chaos. Their mission is pure and simple: to cause trouble.

Bots and trolls often work hand in hand. Automated bots can amplify the impact of trolls through mass engagement, adding thousands of "likes" and comments on content generated by trolls. They are also able to autonomously post content of their own. Bots are created via computer programs, but bot farms (also known as "botnets") and troll farms can have thousands of real people behind the controls. They get paid to pollute conversations on social media.

Bot and troll armies are the flamethrowers often funded by foreign governments like Russia and China to amplify disputes, exacerbate societal divisions, and interfere with democracies.[1] Eye-popping capital is being poured into this arsenal. For instance, a 2023 report by the US State Department[2] and a 2024 report by Ukraine's military intelligence,[3] respectively, found that the Chinese and Russian governments are investing billions of dollars on global disinformation campaigns, with bot and troll armies as key instruments.

Here's their playbook: These bot and troll farms identify preexisting tensions in the United States and other democratic societies. Then they insert themselves into the discourse to intensify conflicts and exacerbate disputes. Instead of advocating a specific political ideology or agenda, these professional trolls typically heighten emotions around contentious topics. The usual suspects—gun control, immigration, police enforcement, public health, and racial divisions. Their aim is to sow distrust in institutions. They seek to blow up the very concept of "truth" itself. By pitting citizens against one another, they employ the classic wartime strategy: divide and conquer. Their ultimate purpose is the disruption and undermining of democracy itself. Ironic, isn't it? Freedom of speech, with sources unchecked, becomes the annihilator.

The impact of bots and trolls on social media is remarkable. One audit found that nearly 50% of President Joe Biden's Twitter/X followers were fake accounts and over 70% of Elon Musk's Twitter/X followers were fake.[4] In the run-up to his Twitter/X takeover, Musk even claimed that 90% of the comments on his Twitter/X posts were from bots.[5] In 2023, Russian government operatives boasted that social networks detect only 1% of their fake accounts.[6]

Bad actors, even in relatively small numbers, can disseminate content via bots and trolls on a vast scale that causes mass disruption. For instance, Stanford researchers examining the role of trolls on social media found that 74% of all conflicts on Reddit were started by just 1% of accounts.[7]

By gaming the engagement-hungry algorithms of mainstream social media giants, bots and trolls can catapult their content to the top of the newsfeeds of millions of users. This has the effect of swaying opinions, manipulating minds, and stirring up mayhem. These misinformation perps even influence elections.

As reported by *MIT Technology Review*,[8] leaked internal reports from Meta revealed that in the month before the 2020 US presidential election, "troll farms reached 140 million Americans." And likely many more in 2024. Shockingly, the report found that 19 of the top 20 Facebook pages for American Christians were actually run by troll farms in Eastern Europe. Additionally, these troll farms ran the largest African American page on Facebook, the second-largest Native American page on Facebook, and the fifth-largest women's page. Leading up to the election, these pages served their followers whatever content the troll farms deemed most inflammatory, best able to divide a house against itself.

In other words, nearly half of the entire US population saw this content not by choice but because Facebook's engagement-seeking algorithms shoved it into their newsfeeds. In the run-up to the 2020 election, these troll farm pages reached 360 million global users every week.[9] Bots don't just spread false or inflammatory information; they also suppress information on real events. US cybersecurity firm Recorded Future reported that hundreds of thousands of coordinated bot accounts were deployed on various social networks worldwide to obscure genuine information about protests happening in China in 2022.[10]

How Bots and Trolls Secretly Fatten Big Tech's Coffers

We can't truly address the problem of bots and trolls without simultaneously addressing the manipulative algorithms of Big Tech's newsfeeds. Those endless threads keep us tuned in, hooked on seeing the latest post, picture, video, laugh, update.

All the Big Tech social media behemoths have revenue models based on targeted advertising. Naturally, greater user engagement/site activity correlates directly to the ability to serve more ads. The result? Ballooning revenue. This revenue scheme incentivizes these networks to take a laissez-faire approach to moderating bots and trolls, particularly the legions that increase their engagement statistics. These nefarious entities, through their outrageous postings, are driving engagement through the roof on mainstream social media, and the dollars keep rolling in.

During Elon Musk's Twitter/X takeover in 2022, Twitter/X relaunched Twitter Blue (now called "X Premium"), adding a new feature allowing any user to get a verification badge simply by paying a monthly fee. Musk claimed via tweet that verifying profiles was "the only way to defeat the bots & trolls."[11] Unfortunately, this move led to a bot/troll stampede of verified accounts impersonating public figures, companies, and organizations, wreaking mayhem on their brands. These accounts have their tweets "rocket to the top of replies, mentions and search" (as per the wording on Twitter/X's website). Talk about influence peddling on the cheap! Hostile governments and other bad actors can now pay for verification and rocket to the top, granting them a whole new level of disruptive power. All the while Twitter/X's bank accounts get fatter and fatter.

As an aside, in November 2022, I gave Musk some unsolicited advice in my *Wall Street Journal* op-ed, "Elon Musk Can Save Twitter—And Democracy."[12] Needless to say, it went unheeded.

Big Tech's Faint-Hearted Battle

Meta, Twitter/X, and all major social media companies have certainly made attempts to fight back against these entities. For instance, they use machine learning algorithms and AI to detect and remove fake accounts, bots, and trolls. They analyze user behavior, posting patterns and other signals to identify suspicious activities. They partner with third-party security firms to help identify, analyze, and address threats, which include botnets, malicious campaigns, and other coordinated activities. And they also work with law enforcement and government agencies to deal with illegal activity. Let's get real though. Their current strategies amount to the world's biggest game of whack-a-mole in a rigged arcade. Why? Remember: Their engagement-based revenue structures are supercharged by incendiary bots and trolls. This payoff incentivizes a lackadaisical approach despite the legionnaires.

AI Supercharges Bot and Troll Armies

AI chatbots like ChatGPT can be incredibly useful and positive tools. The chatbots are also giving bots and trolls superpowers to spread misinformation and toxic content farther, faster, and more sophisticatedly than anything we've seen before. Why is this happening?

First, AI is becoming increasingly adept at generating realistic "deepfakes"—images, video, or audio that imitate a real person's likeness and portray them doing or saying something they didn't do or say. Deepfakes can spread like wildfire on social media, racking up millions of views before being detected for what they are.[13] This duplicity presents an array of chilling consequences, especially during elections.

In 2024, Presidents Biden and Trump, along with Putin, Zelenskyy, and other political leaders worldwide were victims of AI-generated deepfakes falsely impersonating them.[14] Advancing AI tech enables nearly anyone with a laptop or phone to create and spread sophisticated deepfakes. Powerful governments also are behind the deepfake phenomenon. A 2024 report by Microsoft revealed one of the first instances of China's government using AI-generated images and videos, including deepfakes, to target and manipulate voters in the United States, Taiwan, and elsewhere on social media.[15] One AI video depicted the Taiwanese pro-democracy candidate as having multiple mistresses and illegitimate kids. Just a new tool for classic old-school slander—in this case, revealed for what it was.

Second, tools like ChatGPT can be used to create beautifully convincing text that repeats conspiracy theories and misleading information. A shocking 2024 study by researchers at Georgetown and Stanford[16] found that propaganda generated by ChatGPT was deemed "persuasive" by nearly half the people who viewed it.

AI expert Gary Marcus told the BBC in 2023,[17] "People who spread spam around rely on the most gullible people to click on their links, using that spray and pray approach of reaching as many people as possible. But with AI, that squirt gun can become the biggest Super Soaker of all time."

Managing bots and trolls on the social media giants before our current age of AI was already an uphill battle. Now it's like scaling Mount Everest in flip-flops.

Restoration Networking Solutions on the Way

This is where Restoration Networking steps in. While it's impossible to prevent every bot and troll, we can eliminate the vast majority and restore civil discourse and authentic communication on social media. We can drastically limit the reach of bots and trolls, curtail their influence, and cultivate a new, positive, and authentic social media atmosphere.

Restoration Networking's vanquishment recipe sets the table. It's a multipronged approach that you'll enjoy. But let's keep this story moving on here. (For the seven-point plan, see Appendix D.)

Tips for Defanging: What You Can Do Right Now

Here's the good news. Today, we can elevate our game and become savvy at recognizing the discord bots and trolls perpetrate. It's time for us to take a breath and a beat and give humanity the benefit of the doubt. How? By seeing the bot for what it is and does—and ignoring it. All the bot or troll wants to achieve is to hook us into their mindfrick game. You can counterattack with an elegant sidestep.

As AI tech grows increasingly sophisticated, identifying bots will get trickier. But there will always be telltale signs to look for. You don't want to be manipulated by a bot or waste your precious time arguing with one. Let's up our game and ignore the bait. Here are four ways to spot and sidestep a bot on social media:

1. **Check the Profile.**

   If you suspect an account may be a bot, take a look at its profile. If the account was created recently, in the last year or sooner, that's a good clue it may be a bot. Basic bots often won't have a photo, bio, or link in their profiles. Others may use an image stolen from the Web or an account name that was automatically generated. (If their username contains a series of random numbers like a Wi-Fi password, that's a giveaway.) To see if their photo was stolen, you can take a screenshot and then do a reverse image search on Google. (This takes about 10 seconds.) Yep, Google's scraping your data at the same time—as they do any time you google—but it's pretty cool tech if you want to feel like an online detective cracking a case.

2. **Note Odd Timing.**

   You can learn a lot by looking at a profile's posts over time. For instance, if it has been posting at an implausible frequency or during strange hours, often that means it's a bot. Bots can also generate text much faster than humans can type, so if it responds to your replies at lightning speeds, it's probably fake.

3. **Watch for Repetition, Repetition.**

   Bots often serve a specific purpose, leading them to be excessively fixated on particular topics, frequently reposting the same link or posting about the same topic exclusively. If you notice them persistently making the same points, regardless of your input, the account could be fake. A more obvious hint is when they repetitively use the same phrasing.

4. **Stump Them with Non Sequiturs.**

   If you think you might be talking to a bot, often you can throw them off with a non sequitur. Right in the middle of your convo, ask random questions that are completely off-topic: "So, how do you like your eggs?" Bots aren't prepared for this kind of thing. Often they either will awkwardly try to revert to their original topic or will give you something generic, like "Sorry, I don't understand." In those cases, you can bet you're dealing with a bot, sunny-side up. Block and report away.

■ ■ ■

Our conversation about bots and trolls drops us at the footsteps of a kindred yet equally bewildering quandary. How do we—billions of us from around the world—balance on the narrow highwire between our desire for free speech and the necessity of moderation?

17

Balancing Act: Free Speech versus Moderation

Back in Web1, moderation wasn't a big deal. Typically, we were more polite online back then; Miss Manners would've been proud. For the most part, we were interacting with people we knew in real life or with like-minded online communities and hobbyists. There were no armies of bots and trolls fomenting chaos, funded by governments behind the curtains. There were no politicians, pundits, and rabble-rousers slinging digital daggers and arrows across our bows. How times have changed.

In the Web2 world, our newsfeeds are ablaze with fury. Bigotry, prejudice, unsavory language, images, and videos promenade before our eyes.

When it comes to political discourse, Web2 is a mess. A 2023 Pew Research study found that "nearly 80% of those in the U.S. said they believe social media has made people more divided in their political opinions, and roughly 70% said the platforms have made people less civil in how they talk about politics."[1]

As we examined in Chapter 5, currently Web3 offers no respite—all too often its tokenized moderation scheme leads us to even darker echo chambers of doom.

Yet despite these hazards, we must have the freedom to express ourselves without Big Brother's heavy-handed muzzle clamping our tongues. How do we solve this puzzle in a way that satisfies all sides?

Supporting free speech principles and civil discourse is vital to strengthening our republic. Open discussion is what fuels the democratic process. As I say often, disagreement is the backbone of a well-functioning democracy. As American abolitionist and statesman Frederick Douglass said in 1860, "To suppress free speech is a double wrong. It violates the rights of the hearer as well as those of the speaker."[2]

Restoration Networking strives to resolve the core issues regarding content moderation, safeguarding civility while simultaneously supporting free speech tenets. Imagine our collective sigh of relief when this is achieved.

Here's the dilemma we want to resolve. First, sites that forgo moderation altogether in the name of free speech absolutism are dead on arrival. They inevitably become overrun with spam, pornography, hate speech, bullying, harassment, doxxing, and incitement of violence. Not only is this environment dangerous; it also makes these sites unusable for most of us. This has already happened on "anything goes" sites like 8Chan, which was shut down and later relaunched as 8kun. Another site, Secret, was shut down by its founder, who was disgusted by its content. Investors of Secret were refunded their monies.[3]

Second, moderation as doled out by Meta, Twitter/X, and other mainstream giants remains flawed, biased, and untrustworthy, despite ongoing Band-Aids to fix it. A 2024 poll by Pew Research found that 83% of Americans—including majorities of conservatives and liberals alike—believe social media giants "censor political viewpoints they disagree with."[4]

The Six-Point Action Plan for Authentic Civil Discourse

Restoration Networking forges a distinct new path to support authentic civil discourse, a key sweet spot for well-functioning democracies. There are a few simple rules for us as users, enforced by the site's Trust and Safety team: no inciting violence, no bullying, no harassment, no spamming, no doxxing, and no hateful posts. That last rule will be clearly and narrowly defined by the site's User Advisory Board (see Chapter 10) in a way that partisans on both sides of the political aisle can agree with.

These rules allow people of all viewpoints to engage in hearty conversations about politics, health issues, diets, science, and lifestyles. For one (perhaps controversial) example, what this means is that we can always have a lively conversation about the merits of vaccines. It also means that we will always have the freedom to sharply criticize our governments and political leaders. Web and social media companies in China offer a perfect example of where we don't want to go: These companies outright ban all images and mentions of Winnie-the-Pooh, because some Chinese citizens have humorously used the honey-loving bear to mock leader Xi Jinping.[5]

The approach we're taking isn't conservative or liberal. It's American and the foundation of Restoration Networking in the spirit of civil discourse. Consider the following prescription. How does it look to you?

1. No Boosting or Amplification

This book references the key functionality of no boosting or amplification in several places, because eliminating all boosted and amplified content has measurable impacts across the board. It's a key tenet of Restoration Networking. Eliminating boosts and amplified content solves numerous core ailments currently impacting civil discourse on social media. It prevents users from seeing content from entities and individuals they haven't chosen to connect with. No one can amplify or target you with extreme or polarizing content, which is woefully commonplace on Big Tech platforms.

2. Political Neutrality

We all have our own political opinions and biases—that's part of being human! To earn public trust and appeal to a wide user base, the executives and employees of a Restoration Network must avoid letting their politics impact how their site is run.

Here are examples of what *not* to do: Meta deleted hundreds of conservative-leaning political Facebook pages that didn't violate its Terms of Service[6] shortly before a US election. Twitter/X under Elon Musk throttled back traffic to liberal-leaning news outlets such as the *New York Times* by adding delays when Twitter/X users clicked on their links.[7] And as reported by *The Guardian*, TikTok censored videos "that mention Tiananmen Square, Tibetan independence, or the banned religious group Falun Gong."[8]

There are so many examples of these shenanigans, if I listed them all, this book would be longer than *War and Peace*.

Restoration Networks will not engage in such political chicanery. Instead, they will operate under the principle of functional neutrality. Neutrality is paramount to a positive and sustainable social networking experience. To preserve the principle of neutrality, Restoration Networks will have integral systems of checks and balances we'll detail further, including the User Advisory Board, User Juries, Peer Views, and more.

3. Build Trust with Open Source Code

For a variety of reasons, most of us have lost trust in mainstream social media giants, including the platforms' moderation decisions. How do we know if our posts are really reaching our friends or followers? Or if the reach of our posts is being muzzled because of our political/ideological beliefs or backgrounds?

Restoration Networks change this paradigm via engineering with open source code that is publicly available for anyone to inspect and analyze. Maybe you're not savvy about code and algorithms—no problem! Automated algorithms are constantly running in the background of social media sites, determining what content you see, when you see it, and from whom. With the site's open source code, you can see exactly what its automated algorithms are doing with your posts and with your newsfeed. Open source also means AI functionality can be inspected. This transparency is a major step toward regaining trust and accountability from all of us.

4. You Moderate Your Newsfeed, You Decide What You See

Restoration Networks put you, the user, in the driver's seat to decide what kind of experience you want. As we explored in Chapter 8, that includes giving you robust tools to customize your newsfeeds. You are the Feed Curator of your experience.

With the help of AI tools available within the platform, you have the power to curate what kinds of content you see in your newsfeeds—for instance, you can decide if you only want to see content related to certain topics, such as music, fitness, movies, and the like.

Additionally, you can adjust your newsfeeds to automatically hide or provide warning labels for any content you specifically *don't* want to see—for instance, nudity, violence, offensive language, and so on. You have the freedom to adjust the atmosphere and tone of your newsfeeds.

5. User ID Verification

As we've seen, User ID verification has an important role to play—more than one. It is also an excellent structure that reduces onsite moderation. Think of it this way: If nearly everyone you interact with is a real person (with just a sliver of bots/trolls sneaking on), this is a breath of fresh air. Imagine the enjoyment of authentic sharing no matter the subject—lifestyle, music, sports, health, whatever. User ID verification restores true civil discourse and reduces the need to moderate the harassment and hate that comes from anonymous users, bots, and trolls.

6. A Jury of Your Peers

Mistakes happen. No one is perfect, including Trust and Safety teams. Restoration Networks can mitigate this problem by implementing novel jury systems so that wrongfully penalized accounts can "have their day in court" and appeal their cases. Here's a suggested structure: User appeals can be reviewed by a jury of peers. For instance, the social app Minds has used a user-jury system for several years. In each case, the company randomly selects 12 fellow users of the platform to be the jury. Its CEO said, "It's working beautifully."[9] Let me add a caveat here: Minds allows some unsavory posts that wouldn't pass muster on a Restoration Network. However, in principle, this is an interesting idea for the User Advisory Board to contemplate.

Jonathan Zittrain, a professor of law and computer science at Harvard,[10] wrote that a user-based social media jury "brings otherwise intractable conflicts to resolution and legitimacy, even though some people, even many people, will understandably be disappointed by any given decision that emerges from it."

Supporting Section 230 with Careful Reforms

Any conversation about free speech on the Web inevitably intersects with Section 230. I'm passionate about this often-misunderstood legislation. You can find me talking about this in many media outlets;

my favorites are my op-eds in the *Wall Street Journal* in 2021[11] and the *Los Angeles Times* in 2023.[12]

So, what is Section 230 of the Communications Decency Act? It is powerful 1996 legislation that shields Web and social media companies from legal liability for content posted by users. Remember, I am not for "anything goes," and never have been. Yet without Section 230, even comedians like Conan O'Brien[13] probably couldn't get away with tweeting jokes like "I don't need to buy anything on Black Friday, I just want to get in a fistfight at Bed Bath and Beyond to feel alive." Sites like Twitter/X might not feel comfortable hosting this content if they feared there was even a remote chance that Bed Bath & Beyond could sue them.

Section 230 remains a vital piece of legislation to safeguard our free expression on the Web. It also protects small and medium-size companies without deep pockets or armies of lawyers to fend off countless lawsuits that would otherwise drive them out of business. It gives companies broad leeway to moderate their sites at their discretion without liability, and, most important, it enables startups to challenge Big Tech companies in the free market. The Electronic Frontier Foundation has called Section 230 "the most important law protecting internet speech."[14]

Section 230 has drawn fire from both sides of the aisle. Conservative and liberal politicians alike have sought to limit the power that online platforms have over their moderation decisions. Section 230 must be protected, but we can make careful reforms that support free expression, fair competition, and appropriate liability. (For a deeper dive into Section 230 and my legislative recommendations, head to Appendix A.)

18

Facts, Opinions, Lies: Who Decides?

I'VE THOUGHT A lot about fact-checking. Its current versions are imperfect, occasionally biased, and controversial. Do we need to ditch the concept altogether?

Perhaps you've heard the old saying, "A lie can travel halfway around the world while the truth is putting on its shoes." Every day we're bombarded with opinions, facts, and opinions presenting themselves as facts. Discerning the truth can be like finding a needle in a haystack 50 stories high.

The Checkered History of Fact-Checking

In the earlier days of social media, there was no concept of "fact-checking." Maybe our critical thinking skills were better back then because we hadn't been targeted and inundated with countless mistruths. It was common sense to not believe everything you saw or read online.

In 2016, Meta began partnering with third-party fact-checking organizations, including the Poynter Institute's International Fact-Checking Network (IFCN). Meta began to flag and reduce the distribution of certain content, stating it would "identify and address viral misinformation, particularly clear hoaxes that have no basis in fact."[1] Subsequently, in 2022, Google and YouTube provided a $13.2 million grant to the IFCN.[2]

In recent years, social media giants Facebook, Instagram, Twitter/X, YouTube, TikTok, and others have cranked up the dial. They've added fact-check labels, reducing distribution of posts labeled "false," and banned countless users and groups for posting content deemed untrue across broad topics like politics, health, science, and medicine.

Yet fact/truth verification is a nascent movement. A handful of educational and entrepreneurial initiatives are attempting to solve this age-old conundrum of discerning fact from fiction. One of the most prominent organizations in this space is FactCheck.org.[3] It says it relies on unbiased sources of information, nonpartisan government agencies, and experts to "monitor the factual accuracy" of all kinds of statements purported as facts in regard to events, science, medicine, politics, and more.[4]

This and other perceived heavy-handed fact-checking systems have faced controversy from all sides of the political spectrum. Critics argue their checks are insufficient, biased, or applied inconsistently.

True Lies

There have been numerous examples in recent years of social media giants penalizing users and groups for content deemed as "misinformation," later revealed to be valid. This includes news stories about Hunter Biden's laptop in 2020, which resulted in countless users being censored by Meta and Twitter/X, including America's oldest newspaper, the *New York Post*.[5] In another jarring example, countless users of these platforms were censored and penalized for their posts suggesting COVID-19 originated in a lab[6]—despite the FBI and other agencies now deeming that claim to be likely true.[7] People on all sides of the political aisle have been punished by social media giants for posting "false" info that wasn't false at all.

It's a real challenge to discern whether platforms should fact-check someone's statements. After all, by definition, an opinion or perspective is just that. "Rightness" about a position, whether it is related to politics, medicine, health, fitness, science, or anything else, has often been reversed or changed over time.

In 1633, Renaissance scientist Galileo was put under arrest for publishing "misinformation" about the Earth revolving around the Sun, which contradicted the official "truth" that Earth was the center of our universe. If Facebook and Twitter/X had been around back then, Galileo surely would have been booted off their platforms, with large fact-check labels placed over his posts.

The Streisand Effect

Censorship of opinions and purported facts (whether true or false) is often counterproductive and can unintentionally amplify what was meant to be hidden. It's human nature to want to see what's forbidden, which is why books sometimes become more popular after they're banned.

In the internet era, this phenomenon occurs so frequently that it has its own moniker, "the Streisand Effect." This coinage originated in 2003 when the singer Barbra Streisand attempted to have photographs of her Malibu residence removed from the internet.[8] Instead, her efforts caused the photos to go viral. The Streisand Effect demonstrates what happens when we attempt to suppress information or speech. These actions often backfire and result in the very attention we sought to avoid.

When it comes to conspiracy theories on social media, this phenomenon is supercharged. Banning such content often proves to the conspiracy-minded that it must be true and valuable because the authorities don't want them to see it.

A Misleading Warning

You've probably seen fact-check flags on certain social media posts. In your head-scratching moment, what did you decide about it? According to MIT research,[9] when people see that some posts on social media have warning labels, they're far more likely to assume, incorrectly, that all the posts without these warning labels have been verified by fact-checkers. This misperception is exacerbated by the fact that only a tiny fraction of posts with false or unverified information are checked and marked as such. A completely misleading environment arises where people believe that false information has been given the stamp of approval by authorities because it has not been labeled as false. The human mind works in mysterious ways!

Restoration Networking for Information Seekers

We're all information seekers—that's the way of our world today. In contrast to the current social media giants, Restoration Networks have no targeting, no newsfeed manipulation, and no boosted or amplified posts. This fundamental difference prevents any information or opinion—true or false—from being broadly promoted. Instead, users must deliberately seek out information for themselves. They cannot be targeted by others who wish to reach and manipulate their thoughts or opinions

Eliminating all boosted/amplified posts as well as any whiff of purposefully manipulated newsfeeds is the surest way to combat widespread misinformation on social media. This mandate is foundational to all Restoration Networks. It's a simple, effective solution.

A High Bar for "Wrongthink"

Here's the plan. We significantly reduce the number of posts that need to be fact-checked. Regular users like you and me, who are simply having conversations with our friends, family members, and common interest groups on social media, should not face constant interference by fact-checks on our posts. Restoration Networking platforms will fact-check only those posts that reach a relatively high audience threshold and have been flagged by a high percentage of those who viewed them. Choosing what that flag rate is likely will vary site to site. The User Advisory Board can set the triggers it deems make sense. (See Appendix B.) I think a good general bar may be 100,000 impressions and at least a 20% report rate.

Restoration Networks do not play the role of Big Brother by penalizing users or groups for what George Orwell called "wrongthink" in *1984*. A curious fact about *1984*: The book has been banned in certain countries for being "anti-communist" while simultaneously being banned in others for being "pro-communist."[10] Goes to show the absurdity that can happen when we censor ideas and opinions.

Hybrid Reviews

To remain as neutral as possible, Restoration Networks themselves do not participate in fact-checking. Instead, they embrace a new dual fact-check review system with both independent reviews and peer input. Check this out.

Independent Reviews

Half of the Hybrid Review is the Independent Review by a third-party fact-checking organization. There are several of these organizations out there, and some have faced accusations of bias. To promote neutrality, third-party fact-checking organizations used by Restoration Networks will be selected by the users of the network, via the User Advisory Board.

As we discovered in previous chapters, AI is already being used to spread false information rapidly. Now, fact-checking organizations are building their own AI-powered tools to fight back. "Having this technology is a huge step . . . to find more facts to check, and debunk more disinformation," said Irene Larraz, a fact-checker at *Newtral*,[11] a media and data verification company. "Before, we could only do a small part of the work we do today." The other half of Hybrid Reviews are called "Peer Views."

Peer Views

Peer Views take inspiration from Wikipedia's community-driven model. Twitter/X's "Community Notes" feature (formerly known as "Birdwatch") also has produced some promising results. For instance, in November 2022 when Twitter/X's then CEO Elon Musk tweeted that Twitter/X is the "biggest click driver on the Internet by far," a Community Note label from Twitter/X users appeared under his tweet, correcting him: "The reverse is true. Twitter drives 7% of web traffic referrals. Facebook drives 74%." Musk later deleted his tweet.[12]

Here's how Peer Views work: Any post that meets the aforementioned 100k impression/20% flag becomes fact-checkable. This is a way to limit these types of flags and help users assess the verifiability of controversial, high-engagement posts. It is similar to Community Notes with refinements. Users can add context, clarifications, or other types of fact-checks via a tag on the original post.

If a Peer View is rated useful by the community, it is displayed alongside the original post. The nuances of a Restoration Network's Peer View mechanism and any associated algorithms will be decided by its Trust and Safety Team and User Advisory Board. Everything will be transparent and fully revealed to users.

In addition, employees of the Restoration Network cannot submit, edit, or modify Peer Views. This feature is purely user-centric, empowering users to help each other on the platform.

Maybe you're someone who just doesn't want to see fact-checks cluttering up your newsfeeds? No biggie! Here's an additional novel approach that gives you more freedoms than current Big Tech giants allow—*you can choose to opt out of seeing fact-checks in your newsfeeds.*

Critical Thinking Education

In addition to the promising results of MIT's media literacy course we discussed in Chapter 13, a 2023 study by researchers from Michigan State University underscored[13] the power of critical thinking education. The researchers found that educating people to make more discerning judgments about what they see online is a more effective way to combat disinformation than banning or censoring content. The study showed how banning certain content with false information is akin to a Band-Aid that doesn't solve the underlying issues. As reported by psychology news site PsyPost, "Teaching people to recognize their biases, be more open to new opinions, and be skeptical of online information proved the most effective strategy for curbing disinformation."[14]

In this spirit, Restoration Networks are encouraged to provide educational resources for all users to challenge their own biases, recognize disinformation when they see it, and maintain a critical eye when consuming content online.

19

Seven Lessons for the Street Fight

So . . . what have we learned in this book's inquiry? Let's glean the insights that can rocket us forward. The parallels of Big Ag, Big Energy, and Big Tech provide seven bedrock lessons we can apply as we craft a new future for the Web and social media.

Lesson 1: The Myth of Inevitable Dominance

Big Ag, Big Energy, and Big Tech all benefit from the common misperception that their dominance is inevitable—that no upstart alternatives can present a viable challenge.

This notion is absurd on its face. Let's look back at the top 10 biggest companies in America over the past century. Today, most of them of them are either out of business entirely or significantly diminished and usurped by newer players.[1] Companies like Bethlehem Steel, Armour & Co., Swift & Co., and International Harvester were once economic titans—who remembers them today? The same can be said if we look back 50 years. IBM was the most valuable company in the world, and Gulf Oil, Eastman Kodak, and Polaroid were its peers at the top.

Not long ago, regenerative farming was viewed as a niche trend practiced by a handful of small farmers. Today, more than 100 million acres of land worldwide are dedicated to regenerative agriculture, a number expected to increase to 1 billion acres by 2050.[2] Meanwhile, a 2023 report by the International Energy Agency found that solar, wind, and other renewables are expected to surpass coal (the longtime leader) in global electricity production by 2025.[3]

These statistics remind me of the common misperception that social media companies *must* use Surveillance Capitalism and algorithmic manipulation to succeed, grow profits, and maintain mass popularity. This is a myth promoted by the Big Tech companies themselves to justify their own business model.

Web platforms can succeed and earn healthy profits while rejecting these harmful business practices. DuckDuckGo, the privacy-protecting search engine competing with Google, reported over 100 million users and $100 million in revenue in 2023.[4] Mozilla operates in a duality with its nonprofit foundation guiding it along with its for-profit corporate entity. It generates revenue over $500 million annually.[5] Brave, a privacy-protecting browser with 73 million users, had revenue of $260 million in 2023.[6]

Lesson 2: Change Comes from the Outside

The power and wealth of the corporate monopolies are vast and entrenched, and they've worked hard to keep their status quo secured. Big Ag companies like Tyson or Cargill won't transform and reject industrial farming or take the lead toward regenerative farming. Neither will Big Energy companies like ExxonMobil or Duke Energy reject fossil fuels and support distributed solar.

Regenerative farming was started by independent farmers. Then consumers embraced it, and over time the movement achieved critical mass. Only then did major corporations like PepsiCo and General Mills take notice and start making changes as late adopters—not for the philosophy but for the greenbacks. Similarly, the renewable energy movement was spearheaded by independent upstarts like SunPower, First Solar, and SolarCity (later acquired by Tesla). Only later did Big Energy corporations take notice and start adding renewables into their energy portfolios. Interestingly, by doing so, Big Energy actually fanned the flames of these movements, giving the upstarts greater credibility and more business.

In the same vein, a movement that rejects Surveillance Capitalism and advances ethical alternatives won't be led by Big Tech giants like Meta or Google. Social media's transformation requires new people, new platforms, and new paradigms to support pure, user-centric models that respect and empower users.

Lesson 3: Don't Give Up, Embrace Alternatives

If you oppose the practices of factory farms, that doesn't mean you have to give up meat (if you're a meat eater). Regenerative farms with humane animal husbandry, along with plant-based and lab-grown meat startups, are providing healthy, ethical alternatives.

If you're concerned about the practices of Big Energy conglomerates, that doesn't mean you have to go Amish. Solar/wind and other clean-tech companies provide abundant energy and reliable services without harming the planet.

Early adopters change the world. When Elon Musk took over Tesla (yes, he didn't start it),[7] electric cars were still viewed as a niche subset unlikely to become mainstream. Yet enthusiastic early adopters, along with Musk's unflagging perseverance as an industry outsider, led Tesla to become the most valuable auto company on Earth. (As an aside, Musk's takeover of Twitter/X is a completely different scenario. Twitter/X was already a mainstream Big Tech giant when Musk took over, not a startup. The company's fundamental Surveillance Capitalism business model remains intact.)

Most of today's successful Conscious Capitalist companies came from similarly humble beginnings. Patagonia was founded in 1973 by a small group of rock climbers and surfers who shared a love for the environment.[8] Whole Foods started as one small health food store in Austin, Texas, in 1980.[9] Both companies developed loyal customer bases who eventually helped propel them into the massive, global, and highly profitable brands they are today.

If you're fed up with Big Tech, you don't have to unplug from the Web. Instead, become an early adopter of trustworthy, respectful, and fun alternatives. The whole point of Restoration Networking is for all of us to enjoy the remarkable benefits of the Web and social media, without the Big Tech baggage. The moral of the story: Don't give up; embrace or create next-gen alternatives instead.

Lesson 4: Customers Willingly Pay for Healthy Choices

Regenerative farmers like Gabe Brown (mentioned in Chapter 3) are succeeding by marketing directly to customers and providing the trust, transparency, integrity, and nutrient-rich foods that their customers desire. In return, their customers are happy to pay higher prices.

Social media can take a similar approach. Provide users with a healthier experience, and they'll likely be happy to support the social media company in return. Social networks can eliminate the middlemen of advertisers and marketers and instead market directly to customers on a foundation of trust and transparency.

Big Tech, like Big Ag, has traditionally focused solely on quantity (getting as many users as possible as quickly as possible and manipulating their minds to keep their attention for as many hours as possible). Restoration Networks, like regenerative farms, can provide social media users the enriching experience their customers crave. Mainstream users are ready to support social networks that have higher "nutrient density."

Lesson 5: Reward Customers with Dividends

As mentioned in Chapter 3, net metering policies passed by state governments have played a powerful role helping the solar movement shine. Many owners of solar arrays are earning money every month from the energy they generate and sell back to the grid. Solar owners love the sense of ownership, control, and independence they get from generating their own power.

In a similar manner, some B Corps take the commendable step of treating their customers like company shareholders. For instance, REI co-op members receive the REI member dividend,[10] which is a share of REI's profits of the previous year.

Borrowing from these models, Restoration Networks monetarily reward users (not just celebrities and influencers) based on their activity, the engagement of their followers, the content they contribute, and the users' option to monetize their data (with privacy protections). Restoration Networks treat users as a dividend-recipient group who receive a share of the revenue pie.

Lesson 6: Third-Party Certifiers Build Trust

As mentioned in Chapter 3, B Corporations are for-profit companies that have been certified by the nonprofit organization B Lab to meet high standards of social and environmental performance, accountability, and transparency.[11]

B Corps are required to take into account the impact of their decisions on all stakeholders, including workers, suppliers, customers, community, and the environment. They are also required to publicly report their performance using the B Impact Assessment, a standardized tool that measures a company's social and environmental performance.

Restoration Networks can take inspiration from this protocol and develop a parallel program to certify social media companies that demonstrate high ethical standards. In the Restoration Networking universe, its institute will establish both the criteria and the awards/acknowledgments of this program. (See Appendix B.)

Lesson 7: Regulations Work with Real Teeth

As we examined in Chapter 2, Big Ag, Big Energy, and Big Tech companies have faced numerous fines and regulations over the years, often to little effect. However, there have been notable exceptions—powerful regulations that made substantive positive change.

For example, the Superfund program was launched way back in the days of perms and shoulder pads,1980—and it's still an effective regulation to this day.[12] Superfund is distinctive because it requires violating companies to pay to clean up the hazardous waste they create, holding them accountable even after the initial cleanup is completed. Superfund works because it imposes significant penalties—violators can't just pay a nominal fine once and walk away. Bankruptcy doesn't offer a free pass either.

High-impact, enforceable, long-lasting regulations on Big Tech must adopt a similar strategy. The problem is that the government's budgets for regulatory enforcement are far smaller than the budgets for negotiating attorneys deployed by Meta, Google, Twitter/X, and other players. Regulations must mandate minimum level fines that can't be negotiated.

Here are eye-openers that paint this picture. As we examined in Chapter 2, Meta's $5 billion fine by the Federal Trade Commission in 2019 (the biggest fine ever by the FTC against a tech company) could have been over $7 trillion if maximally enforced.[13] Google was fined a "mere" $391 million in 2022 for secretly tracking its users locations after explicitly promising not to.[14] As reported by the BBC, "location services help Google generate $200 billion in annual advertising revenue."[15] Twitter/X was ordered to pay $150 million in 2022 for a "bait-and-switch" scheme in which it asked its users to submit their emails and phone numbers to "improve security." The company then used that data for targeted advertising.[16] As reported by the *Washington Post* at the time, "critics of the tech industry have warned that such fines are toothless against some of the most well-resourced companies in the world."[17]

These companies routinely calculate that the money they make by breaking the rules far exceeds the fines they incur. Another incentive for their incessant bad behavior is that we're not paying attention—they're burying the stories and we're still hanging around.

New regulations on Big Tech must come with real teeth, not brittle dentures. Here's food for thought: Depending on the severity of the violation, the United Kingdom's Online Safety Act[18] imposes jail time for executives and fines up to 10% of a company's gross revenue.

(To learn more about this and other helpful legislation, have a look at Appendix A.)

20

Welcome to Web4: The Restoration of Sanity

RESTORATION NETWORKING REPRESENTS a clear catapult in the evolution of Web iterations. From a nomenclature perspective, these iterations have come to be known sequentially as Web1, Web2, and Web3.

The transformative spirit of Restoration Networking merits a new home. At the time of writing this book in 2024, the term "Web4" has not yet entered mainstream parlance. Some have attempted to define it; for example, the European Commission[1] describes Web4 as a combination of artificial intelligence, the Internet of Things, blockchain, and virtual reality. (As an aside, the Internet of Things is not a topic for Restoring Our Sanity Online. This aging phrase simply references what we're now all accustomed to—the incessant connectedness of most things in our lives to the Web. Doorbells, lights, cameras, washing machines . . . the list is endless.) However, this definition of Web4 is vague and abstract. In fact, it sounds a lot like Web3, which already includes those elements.

It's critical to articulate Web4 as a new iteration of the Web, not another example of "new boss equals old boss."

So, let's put proverbial pen to parchment for Restoration Networking in Web4. Although Web4 will not be perfect, Restoration Networking will be its North Star. Additionally, the core tenets of Restoration Networking can prescribe a more human-centric approach for a variety of companies beyond social media.

The Restoration Networking Constitution

I've done a lot of thinking about how to create some bedrock for the principles espoused within this book. Taking the high ground always requires a march forward with clear guidelines. Let's crystallize the values of this healing, uplifting, regenerative social media ecosystem. Welcome to the essence of our conversation, the Restoration Networking Constitution.

Rather than dogmatic edicts, these constitutional pillars are a recommended framework. They are beacons for social media entrepreneurs, creators, and users. This constitution lays the foundation for healthy, cutting-edge, holistic social networking immersed in Conscious Capitalism for its management and business model.

Here are the 13 golden guidelines for Web4 Restoration Networking that line up perfectly with the seven lessons from Chapter 19.

1. Profit Sharing

As we explored in Chapter 7, Restoration Networking shifts the social media economic model to one that wholeheartedly embraces users as a key stakeholder class. They share in the spoils they help generate. The model draws inspiration from Conscious Capitalism companies that have effectively implemented revenue-sharing models, and it goes a step further.

Through User Awards, all qualifying platform users (not just a minority of celebrities and influencers) reap the financial benefits of their contributions to the network. As specified by its User Advisory Board, a portion of the company's net profits is divided among qualified users each year. These include monetizable daily active users, Premium Subscribers, Content Creators, Influencers, and Page/Group Owners (Chapters 7 and 8). User Awards embody the fundamental principle of acknowledging and rewarding users who actively contribute to the platform's financial success. It's a win-win for users and Restoration Networking companies alike.

Additionally, the model eliminates Web3's linkage of monetary tokens to voting power. Also gone are content moderation decisions in which an elite minority of users who've amassed troves of tokens exert control. Wallets, greenbacks (primary), and cryptocurrencies (secondary) are welcome on Web4, but they're independent of such lords-and-peasants schemes.

What if you want to monetize your data? Users of Restoration Networks will have the option to share their data with the network and its advertising partners. With stringent privacy protections, they'll receive a fair portion of the revenue generated from their participation in a site's ad model revenue.

This model grants users financial benefits while not impinging on genuine relationships, impartial moderation, or unbiased platform policies and features. It elegantly solves the Web3 problems of tokenized economics and gamified schematics.

2. Safeguarding Kids

Restoration Networking takes urgently needed actions to protect our kids. It establishes new rules, guardrails, and guidelines specifically tailored to nurture and protect kids on social media.

Its framework effectively safeguards children under 13 by prohibiting their access to social media entirely. Simultaneously it implements measures that ensure the safety of young people ages 13 to 17 (and adults).

As detailed in Chapter 13, there is a range of actionable solutions to protect kids and young people on social media. These include User ID–based age verification, segregated user experiences based on age, screen time limitations for users under 18, content restrictions to ensure that only age-appropriate content is accessible to users under 18, and much more.

3. User ID Verification

As outlined in previous chapters, Restoration Networking champions User ID verification, mandating that all users authenticate their actual identities in order to use the platform. This prescription tackles various issues plaguing social media, such as preventing propagators of hate, bullies, nefarious governments, lawbreakers, underage users, predators, and hordes of bots and trolls from masquerading on the platform. User ID verification is an indispensable tool for real humans to triumph in this ongoing battle.

Restoration Networks allow users to use pseudonyms on their accounts after they register with User ID verification. This is another level of privacy/vulnerability protection for a user when desired. Pseudonymous accounts would be clearly labeled as such along with the user's real age (e.g., you couldn't be that 50-year-old, discussed earlier, pretending to be a 13-year-old). At the same time, to protect all users, the network's operators would know the real IDs and could take rapid action against abusive accounts.

User ID verification is the only way to truly safeguard kids and young people. It prevents those under the age of 13 from joining and enables age-appropriate content to be served to users under 18. Without User ID, there's no viable way to prevent kids from lying about their birth dates, as millions do every day on the current social media giants.

4. Civil Moderation

Upholding free speech principles on social media is crucial for strengthening democracies worldwide. Open and diverse discussions are the lifeblood of a functioning democratic process. But the landscape of social media discourse is far from democratic or civil. We face a dilemma that requires an urgent resolution.

Platforms that completely forgo moderation in the pursuit of absolute free speech quickly become overrun with unsavory content, such as spam, pornography, hate speech, bullying, harassment, doxxing, and incitement of violence. This abhorrent ambiance not only poses real dangers; it also renders these platforms unusable for most people.

Alternatively, the moderation practices of mainstream giants are often flawed, biased, and untrustworthy.

Restoration Networking embraces fair and balanced moderation without political bias or double standards. A few simple rules are required: No incitement of violence, no bullying, no harassment, no spamming, no doxxing, and no hateful posts—with a clear and specific definition of these that is understandable and agreeable to individuals across the political spectrum.

Upholding commonsense rules of decency fosters a spirit of civil discourse. Within this context, people of all perspectives are allowed to express themselves freely and engage in robust debates on topics such as politics, health issues, diets, science, and lifestyles. This approach isn't conservative or liberal. It's American.

Moderation practices also include Hybrid Reviews, Peer Views, and a Jury of Your Peers. (See Chapters 19 and 20.) Through these practices, Restoration Networking intends to solve all the legacy issues in moderation schemas. It fosters an environment that encourages open dialogue, respects diverse opinions, and builds trust and camaraderie among its users.

5. Open Source Code

A 2023 report by Ipsos revealed that social media is one of the least trusted industries in the world. Only 22% of people believe that social media companies are "trustworthy"—an even lower number than those who believe in the trustworthiness of the oil and gas industry: 23%.[2] Based on our revelatory journey in this book, this is hardly a surprise.

So, what can social media companies do to earn back our trust? As we've explored in previous chapters, a critical component to regaining user trust is open source code.

What's so great about open source? For starters, it allows you and anyone on the Web to take a good peek behind the scenes of a Web company. You can inspect the code and make sure there's no funny business going on. For instance, you can see for yourself if your posts or the posts of your friends or favorite influencers, groups, and pages are being suppressed or muted by any biased AI, algorithm, or company operators.

Throughout the history of the Web, there have been recurring clashes between open source versus closed, proprietary systems. At the beginning, Tim Berners-Lee pioneered the Web's open source nature with a vision for maintaining transparency and trust. Not long after, the Big Tech giants opted for closed systems to conceal how their sausage is made.

Restoration Networks adopt the spirit of the early Web by embracing open source code. This transparency provides users with indisputable evidence that they are not being subjected to unfair manipulation or deception—thus helping to earn users' trust.

6. Vanquishing Bots and Trolls

We've spent time in this book together lifting the veil on the menace of disruptive bots and trolls. Check out Appendix D, the "Seven-Point Prescription to Vanquish Bots and Trolls," for the full vanquishing menu.

But let's not forget that there are "good bots" too. Certain types of bots can provide helpful services for users and are more akin to C3PO than the Terminator.

Along with increased trust and transparency, open source code provides another benefit—directly in this battle. Outside engineers, along with those employed by the company, can contribute and make improvements to the site's code and defenses.

Restoration Networking encourages users who are developers to build and contribute these "good bots" to combat malicious ones in this ongoing battle. As we've examined, the recent explosion of generative AI tech has handed a shiny new flamethrower to bot and troll operators seeking to cause disruption. Restoration Networking's strategy? Fight fire with fire.

As stated by cybersecurity company RSA, "The only way that organizations can keep up with the rate of change is to control bots by using bots: the same underlying principles that allow a ChatGPT to write progressively better jokes or term papers can also train security systems to recognize and respond to suspicious behavior."[3]

With open source code, combined with well-intentioned AI helpers, outside developers are encouraged to submit helpful bots for approval. Restoration Networking companies can reward the developers for their assistance. (See Appendix D.)

7. Protecting Privacy

The people have spoken—and they want data privacy. According to a 2023 report by the International Association of Privacy Professionals, "68% of consumers globally are either somewhat or very concerned about their privacy online."[4] A survey by the *Wall Street Journal* and NBC News found that 74% of Americans believe social media companies collecting users' personal data is not an acceptable trade-off for "free" services.[5] According to an Axios survey,[6] "93% of Americans would switch digital services to a company that prioritizes their data privacy."

The time has come for social media platforms to embrace an ethical business model that places users at the forefront rather than treating personal data as a revenue opportunity for exploitation, sharing, or selling.

On Web4 Restoration Networks, privacy protections are proactively integrated into the underlying technology, policies, and business models of a platform. Privacy is the default setting of Restoration Networks, guaranteeing users absolute control and ownership of their data and content. Restoration Networks reject the creation of data packets on their users, prioritizing user privacy and protection instead of company revenue based on exploitation.

As a user of Restoration Networking, you also have granular control over your privacy settings and permissions. It is you who determines exactly who can access your content and information and who can interact with you or your posts. You also have the freedom to opt out of the search directory and render yourself invisible to other users. These tools can help you prevent bots and trolls from interfering with your experience. This control is critical in safeguarding mental health. Additionally, you can delete your account and take your content with you whenever you choose.

When it comes to data storage, Restoration Networks give users the ability to store and manage their own data, granting them control over their personal information instead of relying on third-party companies (see #10 on this list).

By now you know how I feel about the right to privacy—I've been advocating it for years, ever since Mr. Zuckerberg decided it was bad for his bank account. As I wrote in the *Los Angeles Times*[7] in 2024:

> The United States lags far behind the rest of the world on privacy legislation; 137 of the world's 194 countries have national privacy laws, according to the United Nations. We're the G-20 outlier without one. This isn't the kind of "exceptionalism" Americans should strive for.

Fortunately, the US Congress may be heeding the call. By the time this book is published, I hope that the American Privacy Rights Act will have become the law of the land.

8. Timeline Order

Social media giants like Meta invest exorbitant sums to hire thousands of data scientists and psychologists. Their mission? To manipulate our minds and captivate our attention on their platforms. This insidious approach involves delivering polarizing content that ensures our nonstop engagement and addiction.

Restoration Networking tackles the issue head-on by abolishing all manipulative newsfeed algorithms and adopting timeline-order newsfeeds.

Users can customize their newsfeeds further by filtering out certain keywords or topics or by choosing to selectively view content exclusively from their friends, specific friends, or certain groups or pages.

Unwanted posts from individuals or entities users haven't connected with don't infiltrate newsfeeds—as they incessantly do on the Big Tech platforms. Newsfeeds serve content in chronological order, starting with the most recent. It's a simple, highly effective fix. And a better user experience. Common sense.

The bottom line? Instead of algorithms dictating what you see and how you spend your time on social media, Restoration Networking hands you the reins and puts you in charge. Furthermore, Restoration Networking doesn't stifle or manipulate your reach. Your content reaches everyone you're connected with who wants to see it.

9. Nothing Boosted or Targeted

It's well documented that the engagement-for-profit newsfeed algorithms of the current social media giants intentionally amplify division and extremism. In an interview with the *New York Times*, Twitter/X cofounder Evan Williams described how the algorithms of Twitter/X, Meta, and the other social media giants "reward extremes" and amplify content akin to "car crashes" because it drives higher engagement.[8]

As we've discussed in detail throughout this book, boosted posts amplify the wide range of problems currently endemic on social media. The results range from manipulating our opinions and purchase decisions to fueling a mental health crisis.

Restoration Networks eliminate the ability for any user or entity to boost any post or content. This solution resolves multiple structural issues that undermine healthy discussions on social media. It prevents content from entities and individuals we haven't chosen to connect with and eliminates the widespread dissemination of inflammatory content/falsehoods sourced from malicious actors. This no-boost approach establishes a solid foundation for genuine dialogue and critical thinking.

Restoration Networking brings an end to the era of highly destructive humanity-impairing social media. Sayonara to election interference, authoritarian manipulation, and marketing entities paying social media companies to target and manipulate your thoughts, purchase decisions, and voting preferences. Ad models implemented in Restoration Networking are opt in, revenue sharing, and subject to privacy safeguards.

10. Dual Data Storage with True Privacy

As we discussed in Chapter 7, Restoration Networking will embrace a dual data storage approach that empowers users with freedom and choice. Its networks will have a centralized data repository that upends the Web2 Surveillance Capitalist status quo. Adios to sharing user data with advertisers, marketers, and other third parties for targeted advertising. Also, Restoration Networking supports the optional adoption of Tim Berners-Lee's decentralized Pods initiative, or comparable solutions. As a refresher, Pods (Personal Online Data stores) serve as a private repository for all your data. No peering eyes, no targeting marketers, just you managing who sees what on the networks where you post.

11. Data Portability/Interoperability

Currently, transferring your data and content from a Meta platform to Twitter/X or any other Big Tech platform is impossible. When you switch or join a new platform, you have to start from scratch. No wonder we're all stuck on sites we have irreconcilable differences with. No more!

On Restoration Networks, with the help of solutions like Pods, you have the option to have complete ownership of your content, data, and social connections. You have the freedom to move your data, content, and connections to another platform anytime you choose. The portability of Restoration Networking guarantees genuine freedom for creators while fostering fair competition among social media companies.

To promote the creator economy and free market, Restoration Networking also advocates for new legislative regulations that limit restrictions on content creator mobility, eliminating current anticompetitive barriers. (See Appendix A.)

12. Checks and Balances

How do we liberate ourselves from C-suite tyranny (see Chapter 10)? In short, Restoration Networking embraces a model of checks and balances baked into its core. Regular users are empowered with a voice regarding the platform's direction and directives via the User Advisory Board. And to ensure that a Restoration Network stays true to its principles? That's where the Restoration Networking Institute comes into play.

Here's how we give the institute strong roots. In the agricultural industry, companies can obtain accreditations from reputable third-party organizations to validate their credibility, such as farmers using the "USDA Organic" label. Borrowing from this stamp of legitimacy, the institute grants networks in compliance its official "Certified Restoration Network" designation. This certification lets users know the status and credibility of a Web4 social media platform. (See Appendix B for more details on the User Advisory Board, the institute, and more.)

13. Pay It Forward When Possible

Exemplary Restoration Networks are financially successful and pay it forward by making smart investments in complementary companies.

There are plenty of good examples both in B and C corporations. Here are a few: Patagonia has a venture capital fund that invests in companies working to address environmental and social issues.[9] Unilever has a venture capital arm that invests in startups working on sustainable products and services.[10] Coca-Cola has a venture capital arm that invests in early-stage startups, such as the cold-pressed juice company Suja.[11]

Similarly, Restoration Networks are encouraged to reserve a portion of their profits to fund smaller startups that support the movement and expand the Restoration Networking ecosystem. The startups that receive this funding and the portion of funding they receive can be determined and voted on by the funder site's User Advisory Board. Final approvals are made by the investing company's leadership team.

Bringing It All Together

That's it. These 13 principles offer the framework for the restoration and reinvention of social media and the Web—in its version "4" iteration. These principles are practical, elemental, to the point. No rocket science here, just common sense. Let's get to it!

21

Taking It to the Streets

I WAS THERE at the beginning, and I can tell you firsthand: Social media was *never* meant to be what reigns today. The medium is far astray from its visionary beginnings.

Its North Star was uniting people and providing them with a platform to stay better connected, communicate freely, find like-minded communities, learn, and share their lives. But the experience of social networking careened wildly off course when Big Tech giants started viewing their users as products to spy on, sell, target, and manipulate.

A new dawn is on the horizon. People want social media that reflects reality, not the funhouse mirror distortions we experience on current Big Tech platforms. In the real world, people prefer to lead with kindness, not hatred. They desire common ground, camaraderie, and friendship, not division, hostility, and enmity.

Our Web2 experiences have changed our realities. Let's restore the balance and reclaim civility. We're all human, there's only one spaceship Earth. Entities that aren't necessarily human are destroying our camaraderie. We cannot let them win. Our mission is to reset. A more harmonious future depends on us.

It's encouraging to reflect that even among those with whom we disagree politically, in real life we often get along. I am reminded of the relationship between the late Supreme Court Justices Ruth Bader Ginsburg and Antonin Scalia. Ginsburg was a staunchly left liberal and Scalia a firmly rightwing conservative. You could not find two people whose political views were more opposed. And yet, they were the best of friends. They often went to the opera together and bonded over their mutual love of good wine. Ginsburg even kept a photo in her office of the two of them riding an elephant during a vacation to India.[1]

That's really the whole premise of democracy—the ability to disagree while still being civil, respectful, and affectionate with each other. We deserve social media that supports and nurtures these aspirational fundamentals.

Today, we are participating in the nascent ascension of social media that aligns with the industry's original aspirations: treating users as respected customers to serve and delight.

Friends have often heard me say that "life is perfect in its imperfection." Indeed, I live with this understanding as the fundamental premise to the human contract for life on Earth. But that doesn't let us off the hook. We're not going for perfection. We're going for better, for continuous improvement. The right time to take action is always right now.

The driving force behind every movement combating the amoral giants of an industry—Big Ag, Big Energy, and Big Tech alike—is the commitment to secure a healthier, safer, and empowering future for us, our kids, and future generations. Big Ag, Big Energy, and Big Tech are hulking Goliaths. Undeterred, impressive Davids have emerged in farming and energy. It's time for comparable slingshots taking on Big Tech.

Restoration Networking is a major leap in the right direction. It reshapes the entrenched power dynamics of Big Tech and ushers in a new era. Business revenue and ethical treatment of customers form a potent, harmonious alliance. Social networks can respect their users, achieve mainstream success, and generate significant revenue in a Conscious Capitalism modality.

Now is our time to advance humanity's genuine connectivity. In tandem, we can bolster mental health, personal privacy, civil discourse, and democracy. Starting right now, let's escape from the social media asylum and restore our sanity.

■ ■ ■

This is the end of the book but the beginning of the movement. For all of us, here's what to do next:

- **Give the boot!** To Web2 Surveillance Capitalist companies.
- **Take action!** Join social media companies that share your principles.
- **Talk it up!** Encourage everyone you know to jump on board.
- **Feeling entrepreneurial?** Start a Restoration Network.
- **Share this book:** With friends, family, colleagues, classmates, and others.
- **Contact me:** mark@restoretheweb.com

■ ■ ■

In the pages that follow are appendices that provide extra scaffolding:

Appendix A: Advice to Legislators
Appendix B: Accountability Structures for Restoration Networking
Appendix C: Revenue Superchargers
Appendix D: Seven-Point Prescription to Vanquish Bots and Trolls

Appendix A: Advice to Legislators

Strong legislation must be put in place to protect users. The age-old dilemma remains true: Technology runs far ahead of legislative controls. In my work with all sides and committees, including members of Congress, the Federal Trade Commission, and the US Senate Judiciary, I have spoken about the many pressing need before us. We must protect our kids as well as the mental health of users of all ages. Section 230 can be judiciously updated while carefully supporting startups and the free market. Antitrust regulations require strong enforcement action. The threats of AI necessitate critical oversight and mitigation.

We need legislation passed ASAP—laws with teeth that provide commonsense safety and protections. In this appendix, I map out what I believe are the best and most critical legislative antidotes to remedy and regulate the Web and social media. I believe that all of these are vital.

Protect Kids from the Social Media Abyss

In January 2024, the CEOs of Meta, Twitter/X, TikTok, Snap, and Discord testified at the Senate Judiciary Committee's hearing titled

"Big Tech and the Online Child Sexual Exploitation Crisis." In a rare display of bipartisanship, lawmakers from both parties united in castigating the tech leaders for the harms their platforms perpetrate upon kids.[1] Senator Lindsey Graham went as far as to tell the executives, "You have blood on your hands." Other lawmakers compared the tech platforms to cigarette companies in the way they profit off harmful and addictive products.

Yet the lawmakers bemoaned their lack of progress passing legislation to rein them in. Senator Thom Tillis lamented, "We have an annual flogging every year, and what materially has occurred?"[2]

Let's remember that in May 2023, the US Surgeon General, Dr. Vivek Murthy, issued a dire public advisory[3] warning that social media has a "profound risk of harm to the mental health and well-being of children and adolescents." Dr. Murthy called on lawmakers, tech companies, and parents to "urgently take action."

There always seem to be bills pending that will advance the protections needed. I hope some have passed by the time you read this. Here are seven recommended legislative guideposts to protect kids:

1. **Prohibit Users Under 13**

   We need legislation that is effective in banning and in enforcement of anyone under 13 from social media sites. This includes banning kids from feeder systems like Facebook Messenger Kids, which shamelessly primes children ages 6 to 12 to become lifetime Meta users once they turn 13. There's a misconception that kids under 13 are already prohibited from social media due to the Children's Online Privacy and Protection Act (COPPA) of 1998. This is false. COPPA merely requires that apps and websites get parental consent before collecting information on kids.[4]

2. **Stop Targeting Minors**

   We need to stop social networks from serving all minors (under 18) with algorithmically recommended content. This includes targeting minors with any content, including ads of any kind. Minors, never mind most of us, don't have critical thinking skills developed enough to filter the truths and falsehoods of micro-targeted content designed to persuade and manipulate their thoughts, feelings, and purchases.

3. **No Collecting Personal Data on Minors**

   There must be a no-tolerance policy on collecting the personal information of minors (under 18). This includes information on their friends, contacts, interests, demographics, emotional states, politics, spiritual affiliations, economics, geolocation data, browsing history, biometric data, sexuality, and more. None is to be collected.

4. **Mandate User ID Verification**

   We've got to protect our kids, restore civil discourse, prevent foreign election interference, undermine bots and trolls, and so much more. There's no better way safeguard our future than to recognize today's realities and mandate User ID verification on social media.

   As a privacy advocate, I've previously spoken out against the idea of verification systems. It's worth reiterating that a decade ago—during my time as a Steering Committee Member of the National Strategy for Trusted Identities in Cyberspace (NSTIC), and then with the Identity Ecosystem Steering Group (IDESG)—I argued against the then-proposed national identity system as a violation of individual rights to privacy.

   But it's a brave new world and User ID verification has now become a necessity. The truth is that between the massive data ecosystem and the government's information troves, our identities are already well known. Let's put what they already know to good use in a verification schematic, all the while protecting our privacy every way we can.

   The key to this legislation's mandate is that its sole purpose is to verify the identity of users. It won't track, monitor, or collect data on users. Also, rather than having verification systems solely run within the government's purvey, there should be an option for social media companies to create their own legitimate, regulated verification systems or for new third-party vendors to participate.

5. **Mandatory "Private Mode" for Minors**

   Legislation is required that mandates all minors' profiles on social media are unchangeably set to "private." This is the best solution to prevent them from experiencing unwanted advances from all kinds of bad actors. In this scenario, a minor's profile won't be included in a site's user directories, and no unknown users can tag them, message them, comment on their posts or interact with their profiles. Making a profile private is currently optional on Facebook and Instagram. This must be mandatory for all minors.

6. **Safety Audits for Social Media Companies**

   Through legislation, social media companies must be required to perform annual independent audits to monitor and ensure the safety of their younger users. These companies can conduct and publish annual or bi-annual safety audits by third-party councils on the impact that their site or app has on minors, as well as on the mental health of users of all ages.

7. **Significant Penalties for Violating Companies**

   Tech giants break regulatory rules all the time, disregarding the fines they incur as costs of doing business. Instead, new legislation must impose severe consequences for non-compliance. In this instance, we can learn from the U.K.'s Online Safety Act, which subjects company executives to legal action and potential prison time for failing to adhere to regulations.[5]

Reform Section 230 to Safeguard Fair Competition

Section 230 is a vital piece of legislation. It shields Web companies from liability for content posted by their users. In doing so, it protects small and medium-sized companies from drowning in countless lawsuits. It also gives companies broad leeway for how best to moderate their sites without liability. Most importantly, it enables startups to challenge established Big Tech companies in the free market.

When Section 230 was enacted in 1996, the Web was a vastly different place than it is today. Social media was in utero. Platforms then did not spy on, track, target, or manipulate the online activity of their users. Today, Surveillance Capitalism is the golden goose of all mainstream social media giants. Therein lies the problem: Big Tech companies, including Facebook, Instagram, Twitter/X, TikTok, and YouTube, have abused the privileges of Section 230. They hide behind this legislation's liability shield while targeting their users with content that they did not request or seek out.

Section 230 needs amendments to hold companies accountable while remaining the protector of free market competition. The liability shield it offers must protect a company from lawsuits over content that a Web company does not promote or amplify and from moderation decisions that a company makes that are specifically in line with the company's Terms of Service.

Liability protection can be removed in the following cases:

- Content that a company's algorithms "trend" in front of users who would otherwise not have seen that content.
- Content that has been boosted via a site's paid ad-targeting system.
- Content that has been removed that did not violate any of the site's clearly defined rules for posting that were effective the day it was posted.
- Content that has been recommended or inserted into a user's feed, algorithmically or manually by the site, that the user has not explicitly opted in to. Each company can then make the choice of either being held liable for any targeted content and user manipulation or providing a user-choice operated platform, where newsfeeds only serve content from the sources that the user chooses to see. If this reform took place, algorithmic recommendations would have to become far more transparent on social media platforms. Sites would be required to clearly identify what content was boosted via their algorithms *and* get express permission from their users to serve them that content.

It's best that Section 230 also be amended to require transparency from sites about content moderation policies, as well as to communicate any changes to that policy to their users. This amendment would protect free speech on social media sites because content would be protected from the politically motivated whims of whatever management team runs that platform.

It is also important for Section 230 to identify what boosted content companies won't be liable for. For example, what happens if a social media company recommends a post about big wave surfing and a kid sees the post, goes out surfing, and drowns? Can his family sue the social network?

The solution here is to clarify in the updated 230 legislation that companies are liable for specific types of content they promote, such as libel and incitement to violence, and not just any content that precedes a terrible outcome.

Any broader changes to Section 230 will cause a total loss of user privacy online. If Web companies are held liable for all content on their platforms, they will have to scrutinize everything users post—Big Brother on steroids. Startups do not have the funds needed to afford the monitoring expenses or legal fees.

If Section 230 is revoked, Web companies would have to either censor any remotely controversial content or take a hands-off approach and eschew moderation entirely to avoid liability. The former would be an Orwellian nightmare devoid of free expression, while the latter would mean cesspools of unpalatable content. That is a lose-lose scenario.

Ironically, revoking Section 230 would help Meta, Twitter/X, Google, and other Big Tech social media giants while hurting smaller companies and new startups. The big boys have deep pockets. They can easily hire the massive moderation and legal teams needed to defend themselves. Smaller companies can't. Revoking Section 230 would put hundreds of startups and other smaller companies out of business and entrench Big Tech domination.

Careful reforms to Section 230, however, are needed so that companies are held accountable for the clearly defined content they actively participate in targeting, boosting, or unfairly censoring. Simultaneously, rules need to be set that both ensure the protection of user privacy and avoid frivolous lawsuits. A compromise of Section 230 is the best path forward to ensure company accountability, a healthy free market, free expression, and user privacy.

Respect the First Amendment Rights of Social Media Companies

There are often legislative initiatives seeking to punish social media companies for both the type of content they leave up and the content they choose to take down.[6]

Let's bring some common sense here. Other than content that's clearly lawbreaking, platforms must be free to set their own moderation policies without fearing government retribution. As the late Supreme Court Justice John Paul Stevens acknowledged, "governmental regulation of the content of speech is more likely to interfere with the free exchange of ideas than to encourage it."[7]

Instead of dictating content moderation policies for private companies, lawmakers can focus on strengthening free market fundamentals and fair competition online. All the while supporting authentic civil discourse by minimizing anonymous bots and trolls. In this scenario, if users don't like the moderation policies of Facebook, Instagram, Twitter/X, TikTok, etc., they can seamlessly migrate to viable alternatives. That's the true spirit of free market capitalism.

Antitrust Reforms to Take On Monopolies

In June 2019, I wrote an op-ed in the *Wall Street Journal* titled "I Compete with Facebook, and It's No Monopoly."[8] Fast-forward to August 2021 when the FTC filed an amended antitrust complaint against Meta, which listed MeWe and Snapchat as Facebook's only remaining competitors in personal social networking.[9] Meta responded to the complaint by denying that it's a monopolist and calling the FTC's lawsuit "meritless."[10] Despite MeWe's modest growth, Meta's plethora of anticompetitive actions in the years following my editorial caused me to change my position. In October 2021, two months after the FTC's amended complaint, I published a new op-ed in the *Wall Street Journal*, this time called "I Changed My Mind—Facebook Is a Monopoly."[11]

From years of facing this dilemma head-on (as I write this in May 2024), the case the outcome of FTC's antitrust case against Meta is still pending. Here are recommended regulations to mitigate Meta's monopoly now and in the future:

1. **Decouple Facebook from Instagram, WhatsApp, Threads**
   This is an important first step, though it may be insufficient. They would likely all work in tandem with the same underlying modus operandi. They would continue operating within the same data ecosystem, in which user data is shared with advertisers and marketers across platforms. A Meta breakup may even follow the same destiny as AT&T Corp., which was broken up

by federal regulators into "baby bells" in 1984—only to have many of its parts merge together again over the years.[12] Additional regulations would be required to ensure that the pieces of Meta were completely independent and not collusively or collectively insular.

2. **Protect Content Creators and News Outlets**

   A new approach in antitrust regulation is required to meet this challenge. The independence of content creators and news producers must be protected from Meta's acquisition, quid pro quo, wrath, and retribution. Regulations are needed to protect the independence of news outlets and creators from becoming beholden to a site, as well as from prejudicial retaliatory action if they publish articles negative about Meta, or promote themselves and their content elsewhere.

3. **Mandate Data Portability/Interoperability**

   Meta makes it nearly impossible for their users to take their data from Facebook or Instagram to other social platforms—including their personal photos and posts—because its user data downloader is not formatted to upload the data anywhere else. There must be new legislative mandates for content portability. A universal standard is needed that allows social media users to easily download their data and content from one social network, and then easily upload it to a competitive network if they so choose. This playing field leveler would make it easy for all social media users to move their content, contacts, and fans/followers from site to site.

   This must be universal in its regulatory design. The desired outcome is that content from the social media giants (not just Meta) can be easily downloaded from a previous site and then easily uploaded elsewhere.

   This would help break down monopolistic barriers that currently keep users bound to a particular platform. There have been stalled initiatives attempting to legislate this in the past few years. It's high time to get this done.

4. **Protect Net Neutrality**

   This rule gives all Web companies a level playing field in terms of their access to bandwidth. In other words, a small social media startup is served to its users with the same bandwidth and efficacy as the larger Big Tech apps.

Without it, internet service providers (ISPs) can charge for varying levels of performance. The result of this would be a significant barrier to entry for start-ups with small budgets. Their user experience would be diminished.

This rule has had a seesaw existence. After being established in 2015, it was repealed just 2 years later in 2017. Despite its repeal, a survey found that 86% of Americans supported Net Neutrality. In 2018, California made it a state law. In 2024, it was restored nationally by the Federal Communication Commission (FCC).[13] A key component of fair competition, it is important that Net Neutrality stick around much longer this time around!

Taming AI

This is a Herculean task, and unprecedented because this technology is so unfathomable, from the realm of sci-fi. I took a crack at opening the conversation on regulation in April 2023 when *Fox News* published my op-ed, "Three Ways to Regulate AI Right Now Before It's Too Late."[14] Here are the three:

1. Create a new regulatory commission consisting of AI and ethics experts to establish discreet and enforceable oversight.
2. At the federal level, legislate new regulations requiring full transparency of all AI chatbots and full disclosure regarding which sources AI systems are gathering data from, putting specific rules and guardrails in place.
3. Either voluntarily or through mandate if needed, all content generated by AI chatbots must be identified as such. As reported by *MIT Technology Review*, it is possible to embed hidden "watermarks" into the Large Language Models used by these AI systems, which help computers detect if content was created by AI.[15]

I also tackled AI-generated deepfakes in my March 2024 op-ed in *Fox*, "AI Deepfakes Are Endangering Democracy."[16] As I penned in this piece, "a united front combining technological, legislative, and educational efforts is required." Here are four recommendations I offered:

1. Helpful AI can be used as a tool against harmful AI. It's a learning mechanism. AI can be instructed to become a mastermind detector of deepfakes, identifying subtle patterns humans might not notice. Cross-platform collaboration between major players like OpenAI, Google, Meta, and others to watermark and label AI content is critical and must be strengthened and expanded.
2. There must be strong federal laws established that explicitly protect victims of deepfakes. Even preliminary federal rules and fines could significantly reduce the spread.
3. Social media companies must be held accountable for promoting deepfake content. An effective reform would be to hold companies liable for deepfake content that they play an active role in spreading. This includes via targeted ads or algorithmic boosting, where the content is served to users who otherwise wouldn't have seen it.
4. Improved media literacy is urgently needed. The MIT Center for Advanced Virtuality is a good example that provides online media literacy courses for college students and educators. Similar initiatives can help improve media literacy and critical thinking skills downstream with middle school and high school students.

Coincidentally, subsequent to my 2023 piece, in October 2023, President Biden issued the "Executive Order on Safe, Secure, and Trustworthy Artificial Intelligence."[17] Included are many important parameters and accountability structures for the AI industry. I'm watching with keen eyes to see how the Executive Branch and Congress fulfill the comprehensive requirements within the Executive Order.

Appendix B: Accountability Structures for Restoration Networking

How do we identify, qualify, and then keep Web4 social media companies accountable and trustworthy for their users and the movement's principles? Once again, we can look to the established standards of regenerative farming and renewable energy movements.

Web4 Restoration Networking has four recommended pillars for accountability: Restoration Networking Constitution, Restoration Networking Institute, Restoration Networking Certification, and User Advisory Boards. Each pillar is described in detail in this appendix.

As Web4 Restoration Networking emerges and gains steam, these structures will facilitate credibility and accountability for the movement and the companies involved. So, what are these pillars, and how will they work?

The Restoration Networking Constitution

The Restoration Networking Constitution encompasses the values of the Restoration Networking movement. (See Chapter 20.) Rather than dogmatic edicts, these constitutional pillars are a recommended framework for social media entrepreneurs, creators, and users seeking to build the path toward a better social media future. The constitution lays the foundation for healthy, cutting-edge, holistic social networking immersed in Conscious Capitalism for its management and business model. Here are the 13 golden guidelines:

1. Profit sharing
2. Safeguarding kids
3. User ID verification
4. Civil moderation
5. Open source code
6. Vanquishing bots and trolls
7. Protecting privacy
8. Timeline order
9. Nothing boosted or targeted
10. Data stored with true privacy
11. Data portability/interoperability
12. Checks and balances
13. Pay it forward when possible

The Restoration Networking Institute

The Restoration Networking Institute is modeled after the regenerative movements in agriculture and energy and their independent oversight and certification processes. It is intended to be an independent board of respected technology and ethics leaders. The board provides consultation and certification of new personal social networks as well as other social media companies, inspiring their adoption of Restoration Networking principles.

In the agricultural industry, companies can obtain accreditations from reputable third-party organizations to validate their credibility, such as farmers using the "USDA Organic" label. The Rodale Institute has also developed an official certification for regenerative farming—"Regenerative Organic Certified."[1]

A similar example can be found in the renewable energy sector with the Green-e program,[2] a trusted third-party program that certifies renewable energy and carbon offset products, ensuring they meet the most stringent environmental and consumer protection standards in North America.

Restoration Networking companies can similarly be held accountable by the Restoration Networking Institute. This independent board ensures fair accountability while mitigating bias. It's recommended that the Institute conduct biannual audits of Restoration Networking companies to evaluate their compliance with the approved criteria. Companies that meet the standards would then receive the official Restoration Networking Certification.

If a company is found to be violating the constitution, the Institute can take action in several ways, including and up to revoking the company's official Restoration Networking Certificate.

Restoration Networking Certification

The Restoration Networking Institute determines if a company is founded on and upholds the Restoration Networking Constitution. Companies in compliance receive the official Restoration Networking Certification, which they can display on their sites and apps at their discretion. This certification lets users know the status and credibility of a social media platform.

The Institute can conduct annual assessments to verify that the certified company still adheres to the approved criteria. These reviews are essential in maintaining the credibility of the Restoration Networking Certification. Through a comprehensive and holistic audit, the certification process thoroughly evaluates the entire company, ensuring that it upholds the standards and principles of Web4 Restoration Networking.

Restoration Networking User Advisory Boards

As indicated in Chapter 10, each Restoration Network establishes its own User Advisory Board to ensure fairness and transparency in policies, new features, and User Award distributions. Comprised of users, this board is tasked with making decisions on revenue and profit sharing, which ensures that a company's users get a say in the future of the platform. It also helps achieve a fair balance of power among company owners, users, and stakeholders.

Here's how it can work:

- Any user can fill out a form provided by the company to nominate themselves or someone else (one nomination per person) to be on the User Advisory Board.
- The board is chosen from the list of submissions using either a ranked-choice voting schematic or having the company select from the nominees. This selection process is to be verified by an independent third-party vendor to ensure neither the users nor the company games the system.
- Subsequent elections are held every other year.
- The number of members on the User Advisory Board can vary depending on how many active users are on the platform. Actual size is to be determined by the Institute.
- User Advisory Board members serve two-year terms and cannot be reelected.
- As a form of representative democracy for the broader user base, the User Advisory Board plays a key role in voting on significant decisions, such as policy changes to the platform and decisions regarding allocations of User Awards.
- Product and other decisions (e.g., new features to implement, product roadmaps, other site changes, etc.) are the realms of the employees, board, and executive team. If requested, regular users (not just those on the User Advisory Board) can vote in a form of direct democracy.
- In these instances, users are given a menu of options, preselected by the company, to vote on.
- User who have been active on the platform for at least six months can take part in voting and can vote for new members of the User Advisory Board.

Appendix C: Revenue Superchargers

IN THE REVENUE arena, we've discussed Conscious Capitalism and the Creator Economy (Chapters 7,8). Now let's look at additional revenue superchargers that generate healthy profits and reject Surveillance Capitalism.

Freemium

At a foundational level, Restoration Networks embrace the freemium revenue model. Everything a user needs to have a great experience is free. And there's plenty more in-app for purchase. Examples for users include premium memberships, add-ons, special content, music, NFTs, AI-generation tools, and fan subscriptions. There can also be a variety of freemium features for creators such as creator tools and dedicated stores.

Freemium models work when users love what's free and useful—and are able to fully enjoy a site without spending a penny. There are multiple examples over the years of users who trust and like a platform spending money and supporting the platforms.

Many major companies are already thriving with freemium—consider cloud storage service Dropbox, graphic design tool Canva, and popular music streaming platform Spotify. While Spotify does make a portion of its revenue from ads in its free version, the company reported that over 90% of its revenue comes from its subscriptions, which turn off ads and provide enhanced features.[1]

In the gaming industry, many of the most popular games (and the most profitable companies) embrace the freemium model. Examples include Roblox, Pokémon Go, and Fortnite. The companies earn revenue from in-app purchases for virtual currency, which players use to purchase extra items and accessories in the games.

Ethical Advertising

Advertising via the Surveillance Capitalism model has been Web2's breadwinner for years, the "engine that could" for social media. In 2023 alone, over $600 billion were spent worldwide on digital advertising,[2] pushing ads in front of our eyes everywhere on the Web. This number is expected to increase exponentially in the years ahead.

Where does advertising fit into the healthy, profitable, and Conscious Capitalist social media model?

Some of us feel that ads are enjoyable and helpful to see products and services we may be interested in. On a related note, I spent years researching and patenting a better way to simultaneously personalize and anonymize digital ads to eliminate the "somebody's watching" creepiness.

Here's the fix: Any ad model for Restoration Networking would have to be both "opt-in" and mandated to carefully protect user privacy and anonymity. Users would also have the choice to share their data with the network and its ad partners at their discretion, under careful supervision and privacy rules for engagement. The User Advisory Board can set activity thresholds for users to get a piece of the ad revenue they generate for the site.

Creators would also have an opt-in opportunity to earn a share of ad revenue. For example, if users who have opted in to see ads view an ad displayed during a creator's video, then that creator can earn a portion of the ad revenue. This is like YouTube, Facebook, and Instagram's programs for in-stream ads, but with a distinct and vital difference: Only users who have opted in for ads would see them on the creators' content.

Restoration Networking elegantly solves the problems of Web2 and Web3 with a centralized platform (no engineering foundations in blockchain) and with the option of privacy-preserving ads that reward users with a proportional share of ad revenue. It posits no targeted ads and no mechanisms whatsoever to boost or target anything in a user's feed.

Revenue Sharing with Creators and Page/Group Owners

Revenue-generating activities for creators and page/group owners can include exclusive content; paid subscriptions, livestreaming and live events; exclusive newsletters; coaching and consulting; merchandise; VIP meetups; book and e-book sales; sponsored content; fan clubs; tipping; and more. Page and group owners are encouraged to come up with their own creative ways to generate revenue from their followers/members.

Brands, creators, and influencers can receive additional revenue by having a custom "store" on their page or group. They can create and sell custom emojis, stickers, NFTs, and other virtual products for followers/fans to purchase. Special live-streamed moments/events can be part of the mix. The Restoration Network can receive a slice of each purchase or event—say 20%—in a fair revenue-sharing schematic.

Featured Pages and Groups

Businesses and brands on Restoration Networks can choose to pay for their pages or groups to be included in exclusive "Featured Pages" and "Featured Groups" sections of the platform. This method allows brands to reach larger audiences of users and potential new customers. It doesn't interfere with users' experiences since ads are not disruptively inserted into users' newsfeeds.

Helpful Tools for Creators

Restoration Networks can embrace the model of platforms like Substack,[3] which allows writers to create and distribute newsletters and charge subscribers for exclusive content. Substack provides writers with free tools to manage subscriptions, email delivery, and payment processing. These tools make it easy for writers to monetize their content without having to set up their own infrastructure. The company then takes a percentage of revenue its writers earn. Restoration Networks can take a similar approach, providing all the tools that content creators need to monetize their followers/fans and then taking a slice of revenue from creators' paid subscribers.

Donations

When users value and support a platform, a supplemental donation model can work nicely. It's possible that some Restoration Network founders may prefer a nonprofit model. In those cases, a good example to follow is Wikipedia. One of the most popular sites on the planet with 4.5 billion monthly visits, it is backed entirely by donations from its readers. Wikipedia's average donation is approximately $15.[4] In addition, for-profit Restoration Networks may offer users the option to support them with additional contributions.

Appendix D: Seven-Point Prescription to Vanquish Bots and Trolls

The first step to vanquish bots and trolls is to make a Restoration Network unappealing to them and to the humans/AI pulling their strings. After all, the less they are able to incite or influence users, the less interest they'll have in the site itself. Restoration Networking champions fundamental, structural changes that make vanquishing bots and trolls the priority, not an afterthought. It provides real solutions we can implement right now to dissuade and dismantle the relentless bot and troll advance.

This punch list specifically refers to bot and troll vanquishment. You will see some familiar faces, as they play multiple roles in the schema of a Restoration Network.

1. User ID Verification

As we've explored together in this book, Restoration Networking champions a nontracking User ID verification protocol, requiring social media users to verify their true identities. This is a necessary step to solve several of the thorniest conundrums of social media, including the fight against bots and trolls. Currently bots and trolls exploit the anonymity of social media to spread discord while hiding behind fake or secretive identities. Implementing User ID, where users are both accountable for their posts and protected from bad actors, will help us win this battle.

Restoration Networks make User ID verification the priority while making it as difficult as possible for bots and trolls to create fake accounts that disrupt the authentic communications of real human beings.

2. Eliminate Boosted Posts

Boosted posts contribute to a cluster of problems on social media, from manipulating our opinions and purchase decisions, to spurring a mental health crisis.

The ability to "boost" posts is one of the most powerful weapons in the arsenal of bot and troll purveyors. On all Big Tech social networks, any person or entity can pay to boost a post. In doing so, they reach countless users who otherwise would not have seen that content. Bots and trolls exploit this system to further their advance and manipulate millions of unsuspecting people.

Restoration Networking eliminates boosted posts once and for all—for any user and entity. This solution delivers innumerable benefits, including a major advantage in the discouragement of and fight against bots and trolls.

3. Timeline-Order Newsfeeds

As we've seen, the current newsfeed algorithms of all Big Tech social networks are designed to boost and amplify engaging content. Often this content is what is most divisive and offensive.

Restoration Networking takes the bold solution of eliminating all manipulative newsfeed algorithms. This solution is particularly beneficial in the battle against bots and trolls, besides solving a host of other problems endemic to social media giants.

Instead of newsfeed algorithms designed to serve you intentionally outrageous content to keep you engaged, Restoration Networks have timeline-order newsfeeds with the caveat that you're the only one who can customize your newsfeed.

With the help of AI tools available within the platform, you can customize and curate your newsfeeds for whatever topics or hobbies you like.

That means you simply enjoy posts from the friends, groups, and pages and topics you choose to follow or join. No posts from people or entities you didn't choose to connect with will pop into your newsfeeds, as they do constantly on all current Big Tech networks.

Restoration Networking's timeline approach is straightforward: You'll see posts from either all your friends, groups, and pages in chronological order or from the subset you select. Simple as that.

4. No Targeted Advertising

As we've explored, targeted advertising is one of the core elements of the Surveillance Capitalism business model. This revenue model is the root cause of countless infractions on social media users and the world.

With Restoration Networking, the days of election meddlers, authoritarians, and marketing entities paying social media companies to target users and manipulate their thoughts, purchase decisions, and voting preferences come to an end. On Restoration Networks, there is no way for any bot, troll, person, or business/government entity to pay the company to target you. This approach swiftly hammers the nail in the coffin of the well-funded bots and trolls, who currently are more than happy to pay Big Tech social media companies to target and manipulate their users.

5. Open Source + Allowing Developers to Submit "Good Bots"

Restoration Networks embrace open sourcing both their source code and their algorithmic functions. This approach benefits everyone in two ways.

First, it earns the trust of users while allowing developers and engineers outside of the company to help improve the company's code.

Second, in the war against nefarious bot armies, the moderators of any social network need all the help they can get. Fundamental to Restoration Networking is the active encouragement of and rewards for developers who build helpful bots and AI-enhanced tools that vanquish bad bots and trolls on the site. (Note: These battles may ultimately be hero AI versus villain AI.)

6. Partners and AI Tools for Safety and Security

It's important for Restoration Networks to establish partnerships with third-party organizations that can help keep out bots, trolls, and other bad actors. All Big Tech companies do this.

By teaming up with highly effective partners whose business model is solely to eradicate bad actors, Restoration Networks can better stop illegal activity, bots, trolls, and other bad actors at the door and find them if they've gotten inside.

7. User Participation

Restoration Networks continuously update their users on how to flag, block, and report bots and trolls.

Notes

Introduction

1 **"bad for democracy":** Sneha Gubbala and Sarah Austin, "Majorities in Most Countries Surveyed Say Social Media Is Good for Democracy," Pew Research Center, February 23, 2024. https://www.pewresearch.org/short-reads/2024/02/23/majorities-in-most-countries-surveyed-say-social-media-is-good-for-democracy/.

Chapter 1: My Adventures in Web1 and Web2: The Rise of Surveillance Capitalism

1 **three years in a row:** Kranky Kids, "Kranky Kids Radio Club Interviews." https://www.krankykids.com/info/radio_show/interview_weinstein.html.

2 **participate in this paradigm:** Dave Chaffey, "Global Social Media Statistics Research Summary 2024," *Smart Insights*, February 1, 2024. https://www.smartinsights.com/social-media-marketing/social-media-strategy/new-global-social-media-resea7rch/#:~:text=According%20to%20the%20Datareportal%20January,online%20within%20the%20last%20year.

3 **"No Longer a Social Norm, Says Facebook Founder":** Bobbie Johnson, "Privacy No Longer a Social Norm, Says Facebook Founder," *The Guardian*, January 10, 2010. https://www.theguardian.com/technology/2010/jan/11/facebook-privacy.

4 **off to whoever desires it:** Ashley Ferraro, "What Are Data Brokers?" *Privacy* (blog), August 19, 2022. https://privacy.com/blog/what-are-data-brokers.

5 **collected by the age of 13:** "Your Kids' Apps Are Spying on Them," *Washington Post*, June 9, 2022. https://www.washingtonpost.com/technology/2022/06/09/apps-kids-privacy/.

6 **"more about you than you know about yourself":** Matt Simon, "What This CIA Veteran Learned Helping Facebook with Elections," *WIRED*, July 24, 2019. https://www.wired.com/story/facebook-knows-more-about-you-than-cia/.

7 **"jaywalking and 'frivolous spending":** Nadra Nittle, "Spend 'Frivolously' and Be Penalized Under China's New Social Credit System," *Vox*, November 2, 2018. https://www.vox.com/the-goods/2018/11/2/18057450/china-social-credit-score-spend-frivolously-video-games.

8 **sharing satirical memes:** Jessie Yeung and Yong Xiong, "Man Detained for 9 Days in China for Sending Meme Deemed 'Insulting' to Police," CNN, November 2, 2021. https://www.cnn.com/2021/11/02/china/china-man-detained-meme-intl-hnk-scli/index.html.

9 **inventor of the Web:** W3.org, "Tim Berners-Lee Biography." https://www.w3.org/People/Berners-Lee/.

10 **cofounder of the Conscious Capitalism movement:** Conscious Capitalism, "Raj Sisodia." https://www.consciouscapitalism.org/people/raj-sisodia.

11 **leader in tech ethics:** Sherry Turkle, "Sherry Turkle." https://www.sherryturkle.com.

12 **cofounder of Apple:** William L. Hosch, "Steve Wozniak," *Britannica*,. https://www.britannica.com/biography/Stephen-Gary-Wozniak.

13 **Privacy Bill of Rights:** MeWe, "MeWe Privacy Policy." https://mewe.com/cms/privacy.

14 **amended antitrust complaint:** Federal Trade Commission, "FTC Alleges Facebook Resorted to Illegal Buy-or-Bury Scheme to Crush Competition After String of Failed Attempts to Innovate," August 19, 2021. https://www.ftc.gov/news-events/news/press-releases/2021/08/ftc-alleges-facebook-resorted-illegal-buy-or-bury-scheme-crush-competition-after-string-failed.

Chapter 2: Uncanny Parallels: Big Ag, Big Energy, Big Tech

1 **90% of Americans:** Statista, "Social Media Usage in the United States—Statistics & Facts," September 20, 2023. https://www.statista.com/topics/3196/social-media-usage-in-the-united-states/#topicOverview.

2 **entire human population:** Agence France Press, "More Than 60% of World Now on Social Media: Study," *Barron's*, July 20, 2023. https://www.barrons.com/news/more-than-60-of-world-now-on-social-media-study-516c0d81.

3 **("Big Food"):** Kenny Torella, "Now Is the Best Time in Human History to Be Alive (Unless You're an Animal)," Vox, October 13, 2022. https://www.vox.com/future-perfect/2022/9/12/23339898/global-meat-production-forecast-factory-farming-animal-welfare-human-progress.

4 **lighting for thousands for years:** Department of Natural Resources, State of Louisiana, "How Ancient People and People Before the Time of Oil Wells Used Petroleum." https://www.dnr.louisiana.gov/assets/TAD/education/BGBB/2/ancient_use.html.

5 **drinking water to villages:** "Water Wheel," *Britannica*. https://www.britannica.com/technology/waterwheel-engineering.

6 **assist with food production:** US Energy Information Administration, "Wind Explained: History of Wind Power." https://www.eia.gov/energyexplained/wind/history-of-wind-power.php.

7 **New York in 1821:** American Public Gas Association, "A Brief History of Natural Gas." https://www.apga.org/apgamainsite/aboutus/facts/history-of-natural-gas.

8 **Pennsylvania in 1859:** History of Information, "The First Successful Oil Well Is Drilled in Titusville, Pennsylvania." https://www.historyofinformation.com/detail.php?entryid=3061.

9 **rooftop solar array in New York:** Glen Meyers, "Photovoltaic Dreaming 1875–1905. First Attempts at Commercializing PV," CleanTechnica, 2015. https://cleantechnica.com/2014/12/31/photovoltaic-dreaming-first-attempts-commercializing-pv/.

10 **wind and geothermal energy:** Jared Wolf, "How Nikola Tesla Embodied Sustainability in the Early 1900s," *Sustainable Review*, February 6, 2023. https://sustainablereview.com/nikola-tesla-sustainability/#:~:text=Tesla%20proposed%20electric%20power%20generation,years%20before%20we%20discovered%20it.

11 **the invention of the printing press in 1436:** Dave Roos, "7 Ways the Printing Press Changed the World," *History*, March 27, 2023. https://www.history.com/news/printing-press-renaissance#.

12 **to be available to all, free forever:** Mark Fischetti, "The World Wide Web Became Free 20 Years Ago Today," *Scientific American*, April 30, 2013. https://blogs.scientificamerican.com/observations/the-world-wide-web-became-free-20-years-ago-today/.

13 **knighted by Queen Elizabeth II in 2004:** ERCIM News, "Tim Berners-Lee Knighted by Her Majesty Queen Elizabeth II," October 2004. https://www.ercim.eu/publication/Ercim_News/enw59/w3c-1.html#:

~:text=Tim%20Berners%2DLee%2C%20the%20inventor,on%20Friday%2C%2016%20July%202004.

14 **production from about 40% to over 80%:** Dan Kaufman, "Is It Time to Break Up Big Ag?" *The New Yorker,* August 17, 2021. https://www.newyorker.com/news/dispatch/is-it-time-to-break-up-big-ag.

15 **54% of all poultry production in America:** Ahmari Anthony, "The Meat Industry's Middlemen Are Starving Families and Farmers," *The American Prospect*, February 10, 2022. https://prospect.org/economy/meat-industrys-middlemen-starving-families-and-farmers/.

16 **over 60% in today:** MMR, "Seeds Market: Global Industry Analysis and Forecast (2023–2029)." https://www.maximizemarketresearch.com/market-report/global-seeds-market/111623/.

17 **"this situation is urgent":** Paige Sutherland and Meghna Chakrabarti, "More Than Money: The Monopoly on Meat," WBUR, February 14, 2022. https://www.wbur.org/onpoint/2022/02/14/more-than-money-monopoly-and-meat-processing.

18 **90% of all oil production in America:** Library of Congress, "Standard Oil Established." https://guides.loc.gov/this-month-in-business-history/january/standard-oil-established.

19 **(BP) acquired fellow oil giant Amoco:** Youssef Ibrahim, "British Petroleum Is Buying Amoco in $48.2 Billion Deal," *New York Times*, August 12, 1998. https://www.nytimes.com/1998/08/12/business/british-petroleum-is-buying-amoco-in-48.2-billion-deal.html.

20 **Exxon and Mobil merged to form ExxonMobil:** ExxonMobil, "Our History." https://corporate.exxonmobil.com/who-we-are/our-global-organization/our-history.

21 **Chevron and Texaco merged to form ChevronTexaco:** Federal Trade Commission, "FTC Consent Agreement Allows the Merger of Chevron Corp. and Texaco Inc., Preserves Market Competition," September 7, 2001. https://www.ftc.gov/news-events/news/press-releases/2001/09/ftc-consent-agreement-allows-merger-chevron-corp-texaco-inc-preserves-market-competition.

22 **utility companies were handed legal monopolies:** David Roberts, "Power Utilities Are Built for the 20th Century. That's Why They're Flailing in the 21st," Vox, September 9, 2015. https://www.vox.com/2015/9/9/9287719/utilities-monopoly.

23 **"The Facebook":** History.com, "This Day in History: February 4." https://www.history.com/this-day-in-history/facebook-launches-mark-zuckerberg.

24 **hired gun—wunderkind coder Mark Zuckerberg:** Abhishek Chakraborty, "This Indian-Origin Man Sued Mark Zuckerberg for 'Stealing' His Idea, Got Millions in Settlement," NDTV, April 17,

2018. https://www.ndtv.com/indians-abroad/this-indian-origin-man-sued-mark-zuckerberg-for-stealing-his-idea-1838568.

25 **across its platforms–nearly half the global population:** Daniel Shvartsman, "Facebook: The Leading Social Platform of Our Times," *Investing*, November 23, 2023. https://www.investing.com/academy/statistics/facebook-meta-facts/#:~:text=More%20than%2077%25%20of%20Internet,at%20least%20one%20Meta%20platform.

26 **the tech landscape:** Brian Keogh, "50 Attorneys General Investigate Google," *Transnational Law & Contemporary Problems*, October 27, 2021. https://tlcp.law.uiowa.edu/sites/tlcp.law.uiowa.edu/files/8._keogh.pdf.

27 **worldwide are conducted using Google's platform:** Meg Prater, "31 Google Search Statistics to Bookmark ASAP," HubSpot, August 30, 2023. https://blog.hubspot.com/marketing/google-search-statistics.

28 **for a total price tag of over $40 billion:** Tracxn, "Overview of Acquisitions by Google," 2024. https://tracxn.com/d/acquisitions/acquisitions-by-google/__8zKOTB9XR934x_3BSnEVSrmsu3RZtEv3AorQtuBb2Yk#:~:text=Google%20has%20made%20260%20acquisitions,Consumer%20Digital%20%2D%20US%20and%20others.

29 **YouTube in 2006 for $1.65 billion:** Andrew Ross Sorkin and Jeremy W. Peters, "Google to Acquire YouTube for $1.65 Billion," *New York Times*, October 9, 2006. https://www.nytimes.com/2006/10/09/business/09cnd-deal.html.

30 **40% of the American online retail market:** Sara Lebow, "Amazon Will Capture Nearly 40% of the US Ecommerce Market," *Insider Intelligence*, March 23, 2022. https://www.insiderintelligence.com/content/amazon-us-ecommerce-market.

31 **Walmart, owns a mere 6%, by comparison:** Statista, "Market Share of Leading Retail E-Commerce Companies in the United States in 2023," April 2023. https://www.statista.com/statistics/274255/market-share-of-the-leading-retailers-in-us-e-commerce/.

32 **(AWS) controls 33% of this entire market:** Paul Kirvan, "Amazon Web Services (AWS)," *TechTarget*. https://www.techtarget.com/searchaws/definition/Amazon-Web-Services#:~:text=As%20of%20the%20first%20quarter,Verizon.

33 **Google Cloud is in third place with 8%:** Neal Weinberg, "AWS, Google Cloud, and Azure: How Their Security Features Compare," CSO Online, August 8, 2022. https://www.csoonline.com/article/570833/aws-google-cloud-platform-and-azure-how-their-security-features-compare.html.

34 **Meta has raised over $24 billion in funding:** Crunchbase, "Meta." https://www.crunchbase.com/organization/facebook/company_financials.

35 **$16 billion raised from its IPO in 2012:** Robert Hof, "It's Official! Facebook Raises $16 Billion in Historic IPO," *Forbes*, May 17, 2012. https://www.forbes.com/sites/roberthof/2012/05/17/its-official-facebook-raises-16-billion-in-historic-ipo/?sh=7c1d0f1d30bd.
36 **billion raised from its IPO in 2013:** Jonathan Vanian, "Twitter Is Now Owned by Elon Musk — Here's a Brief History from the App's Founding in 2006 to the Present," CNBC, October 30, 2022. https://www.cnbc.com/2022/10/29/a-brief-history-of-twitter-from-its-founding-in-2006-to-musk-takeover.html#:~:text=A%20decade%20ago%2C%20Twitter%27s%20future,is%20back%20in%20private%20hands.
37 **Cargill earned $177 billion:** Karl Plume, "Cargill Fiscal 2023 Revenue Rises 7% to Record $177 Billion," Yahoo! Finance, August 4, 2023. https://finance.yahoo.com/news/cargill-fiscal-2023-revenue-rises-180900053.html.
38 **JBS earned over $72 billion:** Ryan McCarthy, "JBS Details 2024 Response to 2023 Results," *Supermarket Perimeter*, March 28, 2023. https://www.supermarketperimeter.com/articles/11021-jbs-details-2024-response-to-2023-results.
39 **Tyson Foods earned over $52 billion:** Tyson Foods, "Tyson Foods Reports Fourth Quarter and Fiscal 2023 Results," November 13, 2023. https://www.tysonfoods.com/news/news-releases/2023/8/tyson-foods-reports-fourth-quarter-2023-results.
40 **revenues over $344 billion:** Exxon Mobil, "ExxonMobil Announces 2023 Results," February 2, 2024. https://investor.exxonmobil.com/news-events/press-releases/detail/1156/exxonmobil-announces-2023-results.
41 **Shell over $323 billion:** Shell PLC, "Shell Plc 4th Quarter 2023 and Full Year Unaudited Results," Yahoo! Finance, February 1, 2024. https://finance.yahoo.com/news/shell-plc-4th-quarter-2023-070000120.html.
42 **Chevron over $200 billion:** Chevron, "2023 Financial Highlights." https://www.chevron.com/-/media/shared-media/documents/2023-chevron-annual-report-supplement.pdf#:~:text=URL%3A%20https%3A%2F%2Fwww.chevron.com%2F.
43 **BP over $213 billion:** Macrotrends, "BP Revenue 2010–2023 | BP." https://www.macrotrends.net/stocks/charts/BP/bp/revenue#:~:text=BP%20revenue%20for%20the%20twelve,a%2051.58%25%20increase%20from%202021.
44 **revenues over $29 billion:** Macrotrends, "Duke Energy Revenue 2010–2023 | DUK." https://www.macrotrends.net/stocks/charts/DUK/duke-energy/revenue.
45 **PG&E over $24 billion:** Macrotrends, "Pacific Gas & Electric Revenue 2010–2023 | PCG." https://www.macrotrends.net/stocks/charts/PCG/pacific-gas-electric/revenue.

46 **and Exelon over 21 billion:** Yahoo! Finance, "Exelon Full Year 2023 Earnings: Revenues Beat Expectations, EPS in Line," February 23, 2024. https://finance.yahoo.com/news/exelon-full-2023-earnings-revenues-110020650.html#:~:text=Exelon%20(NASDAQ%3AEXC)%20Full%20Year%202023%20Results&text=Revenue%3A%20US%2421.7b%20(up,in%20line%20with%20FY%202022.

47 **company's faulty and unmaintained equipment:** Jorge L. Ortiz, "'Deeply Sorry' PG&E Takes Blame for California's Deadliest Wildfire, Seeks 'Technologies' to Limit Future Risks," *USA Today*, December 3, 2019. https://www.usatoday.com/story/news/nation/2019/12/03/pg-e-camp-fire-california-regulators/2602300001/.

48 **$5 billion in dividends to its shareholders:** Garrett Hering, "Judge Asks PG&E to Explain $5 Billion in Dividends, Political Spending Amid Grid Neglect," S&P Global, July 11, 2019. https://www.spglobal.com/commodityinsights/en/market-insights/latest-news/electric-power/071119-judge-asks-pgampe-to-explain-5-billion-in-dividends-political-spending-amid-grid-neglect.

49 **"take a corporation and put it into prison":** Jane McMullen, "Fire in Paradise," *Frontline*, October 29, 2021. https://www.pbs.org/wgbh/frontline/documentary/fire-in-paradise/?utm_source=Twitter&utm_medium=Twitter&utm_term=PGE&utm_content=Video&utm_source=Twitter&utm_medium=Twitter&utm_term=PGE&utm_content=Video.

50 **Meta:** Stacy Jo Dixon, "Annual Revenue and Net Income Generated by Meta Platforms from 2007 to 2023," Statista, March 4, 2024. https://www.statista.com/statistics/277229/facebooks-annual-revenue-and-net-income/#:~:text=In%202023%2C%20Meta%20Platforms%20generated,119%20billion%20USD%20in%202022.

51 **Alphabet:** Macrotrends, "Alphabet Revenue 2010–2023." https://www.macrotrends.net/stocks/charts/GOOG/alphabet/revenue#:~:text=Alphabet%20annual%20revenue%20for%202023,a%2041.15%25%20increase%20from%202020.

52 **Apple:** Kif Leswing, "Apple's Stock Underperformed Top Tech Peers in 2023 Due to Longest Revenue Slide in 22 Years," CNBC, December 29, 2023. https://www.cnbc.com/2023/12/29/apple-underperformed-mega-cap-peers-in-2023-due-to-revenue-slide.html#:~:text=Despite%20its%20struggles%2C%20Apple%20remains,%2497%20billion%20in%20net%20income.

53 **Amazon:** Macrotrends, "Amazon Revenue 2010–2023 | AMZN." https://www.macrotrends.net/stocks/charts/AMZN/amazon/revenue#:~:text=Amazon%20annual%20revenue%20for%202023,a%2021.7%25%20increase%20from%202020.

54 **Microsoft:** Lionel Sujay Vailshery, "Microsoft's Net Income from 2002 to 2023," Statista, March 6, 2024. https://www.statista.com/statistics/267808/net-income-of-microsoft-since-2002/#:~:text=In%20the%20fiscal%20year%202023,at%20212%20billion%20U.S.%20dollars.

55 **Today, they account for less than a quarter:** USDA, "Small Farms, Big Differences," May 18, 2010. https://www.usda.gov/media/blog/2010/05/18/small-farms-big-differences.

56 **significant policy barriers:** Jillian Ambrose, "US Oil Giants Top List of Lobby Offenders Holding Back Climate Action," *The Guardian*, November 3, 2021. https://www.theguardian.com/business/2021/nov/04/us-oil-giants-top-list-lobby-offenders-exxonmobile-chevron-toyota.

57 **that would help the solar industry:** Michelle Lewis, "EGEB: Oil and Gas Lobbyists Are Trying to Stop Clean Energy with Facebook Ads," Electrek, September 30, 2021. https://electrek.co/2021/09/30/egeb-oil-and-gas-lobbyists-are-trying-to-stop-clean-energy-with-facebook-ads/.

58 **2012 for $1 billion:** Evelyn M. Rusli, "Facebook Buys Instagram for $1 Billion," *New York Times*, April 9, 2012. https://archive.nytimes.com/dealbook.nytimes.com/2012/04/09/facebook-buys-instagram-for-1-billion/.

59 **WhatsApp in 2014 for $16 billion:** Nick Woodard, "When Did Facebook Buy WhatsApp & How Much Did It Pay?" Screen Rant, December 11, 2020. https://screenrant.com/facebook-whatsapp-purchase-date-cost/.

60 **snap up Snapchat:** CNN Business, "In 2013, Snapchat Rejected Facebook's $3 Billion Bid." https://www.cnn.com/videos/business/2022/10/20/snapchat-facebook-buyout-offer-2013-vault-orig-ht.cnn-business.

61 **and Twitter/X:** Matthew Panzarino, "The Three Reasons Twitter Didn't Sell to Facebook," *TechCrunch*, November 4, 2013. https://techcrunch.com/2013/11/04/the-three-reasons-twitter-didnt-sell-to-facebook/?guccounter=1.

62 **Meta has acquired over 100 companies:** Asia Martin, "Meta Can't Make Any Big Acquisitions Right Now, as Its $40 Billion War Chest Starts Going Towards Appeasing Wall Street Investors," *Business Insider*, March 17, 2023. https://www.businessinsider.com/meta-40b-war-chest-wall-street-metaverse-layoffs-zuckerberg-2023-3.

63 **a dossier to the Federal Trade Commission:** Georgia Wells, "Snap Detailed Facebook's Aggressive Tactics in 'Project Voldemort' Dossier," *Wall Street Journal*, September 24, 2019. https://www.wsj.com/articles/snap-detailed-facebooks-aggressive-tactics-in-project-voldemort-dossier-11569236404.

64 "copy, acquire, or kill." Amanda Bronstad, "Facebook Hit with Antitrust Class Action Over 'Copy, Acquire, Kill' Methods," Law .com, December 4, 2020. https://www.law.com/therecorder/2020/12/04/facebook-hit-with-antitrust-class-action-over-copy-acquire-kill-methods/.

65 **(FTC) filed an amended complaint:** Federal Trade Commission, "FTC Alleges Facebook Resorted to Illegal Buy-or-Bury Scheme to Crush Competition After String of Failed Attempts to Innovate," August 19, 2021. https://www.ftc.gov/news-events/news/press-releases/2021/08/ftc-alleges-facebook-resorted-illegal-buy-or-bury-scheme-crush-competition-after-string-failed.

66 **engine on most Web browsers and phones:** Mike Scarcella, "Explainer: Why Is the US Suing Google for Antitrust Violations?" Reuters, September 11, 2023. https://www.reuters.com/legal/why-is-us-suing-google-antitrust-violations-2023-09-11/.

67 **compared to other video player competitors:** Christina Tabacco, "Google Answers Complaint Against YouTube Competitor Rumble," Law Street, September 13, 2022. https://lawstreetmedia.com/news/tech/google-answers-complaint-against-youtube-competitor-rumble/.

68 **later reversed after swift backlash:** Matt O'Brien and Brian P. D. Hannon, "Twitter Deletes New Policy Banning 'Free Promotion' of Rival Social Media Platforms," *USA Today*, December 18, 2022. https://www.usatoday.com/story/tech/2022/12/18/twitter-bans-links-facebook-instagram-mastodon-accounts/10921759002/.

69 **Reuters, Threads, Substack, and others:** Sheila Dang, "Musk's X Delays Access to Content on Reuters, NY Times, Social Media Rivals," Reuters, August 15, 2023. https://www.reuters.com/business/media-telecom/musks-x-delays-access-content-reuters-ny-times-social-media-rivals-2023-08-16/.

70 **lobbied for new laws:** Nancy Fink Huehnergarth, "Big Agriculture Bullies and Lobbies to Keep Americans in the Dark," *Forbes*, May 5, 2016. https://www.forbes.com/sites/nancyhuehnergarth/2016/05/05/big-ag-bullies-and-lobbies-to-keep-americans-in-the-dark/?sh=415fa3da502c.

71 **organic food as an "expensive scam":** Stacy Malkan, "Monsanto's Fingerprints All Over Newsweek's Opinion Piece, a Hit on Organic Food," EcoWatch, January 24, 2018. https://www.ecowatch.com/monsanto-propaganda-newsweek-2528277875.html.

72 **loaded with antibiotics:** NBC News, "USDA Says Tyson Used Antibiotics on Chicken," June 3, 2008. https://www.nbcnews.com/id/wbna24956860.

73 **into the 2000s:** Tony Briscoe, "ExxonMobil Publicly Denied Global Warming for Years but Quietly Predicted It," *Los Angeles Times*, January 12, 2023. https://www.latimes.com/environment/story/2023-01-12/exxonmobil-accurately-predicted-effects-of-global-warming.

74 **reportedly spent over $30 million:** Shannon Hall, "Exxon Knew About Climate Change Almost 40 Years Ago," *Scientific American*, October 26, 2015. https://www.scientificamerican.com/article/exxon-knew-about-climate-change-almost-40-years-ago/.

75 **hired the same researchers and consultants:** Benjamin Hulac and Climatewire, "Tobacco and Oil Industries Used Same Researchers to Sway Public," *Scientific American*, July 20, 2016. https://www.scientificamerican.com/article/tobacco-and-oil-industries-used-same-researchers-to-sway-public1/.

76 **A peer-reviewed study in 2022:** Amy Westervelt, "The Great Greenwashing Scam: PR Firms Face Reckoning After Spinning for Big Oil," *The Guardian*, February 18, 2022. https://www.theguardian.com/environment/2022/feb/18/greenwashing-pr-advertising-oil-firms-exxon-chevron-shell-bp.

77 **found that PG&E had falsified its records:** Christina Maxouris, "California Commission Finds PG&E Falsified Records for Years," CNN, December 15, 2018. https://www.cnn.com/2018/12/15/us/pge-falsifying-records/index.html.

78 **our newsfeeds during US elections:** Salvador Rodriguez, "Facing Scrutiny, Facebook Reportedly Hired a PR Firm That Wrote Negative Articles About Rivals Apple, Google," CNBC, November 15, 2018. https://www.cnbc.com/2018/11/14/facebook-hired-pr-firm-that-wrote-negative-articles-about-rivals-nyt.html.

79 **don't collect our data from Gmail:** Shoshana Wodinsky, "The Three Biggest Lies Google and Facebook Spouted About Your Privacy Before Congress," Gizmodo, July 31, 2020. https://gizmodo.com/the-three-biggest-lies-google-and-facebook-spouted-abou-1844557693.

80 **from these tech giants:** John Chachas, "Google, Facebook and Twitter Are Killing the Local Newspapers Citizens Need for Democracy," *Dallas Morning News*, February 28, 2021. https://www.dallasnews.com/opinion/commentary/2021/02/28/google-facebook-and-twitter-are-killing-the-local-newspapers-citizens-need-for-democracy/.

81 **Over 70% of Americans:** Andrew Hutchinson, "New Research Shows that 71% of Americans Now Get News Content via Social Platforms," Social Media Today, January 12, 2021. https://www.socialmediatoday.com/news/new-research-shows-that-71-of-americans-now-get-news-content-via-social-pl/593255/.

82 **Tech journalist Jacob Silverman noted:** Jacob Silverman, "Spies, Lies, and Stonewalling: What It's Like to Report on Facebook," *Columbia Journalism Review*, July 1, 2020. https://www.cjr.org/special_report/reporting-on-facebook.php.

83 **advisor and attorney:** David Dayen, "From Google Payroll to Government and Back Again," The Intercept, January 13, 2016. https://theintercept.com/2016/01/13/from-google-payroll-to-government-and-back-to-google-again/.

84 **Big Ag has spent $2.5 billion on lobbying:** Robert Semple, "America Has a Chance to Make Farming More Climate Friendly," *New York Times*, June 12, 2023. https://www.nytimes.com/2023/06/12/opinion/climate-change-farm-bill-farmers.html.

85 **the industry spent over $177 million:** Open Secrets, "Agribusiness Lobbying." https://www.opensecrets.org/industries/lobbying?ind=A.

86 **$750 million to political candidates:** Georgina Gustin, "Big Meat and Dairy Companies Have Spent Millions Lobbying Against Climate Action, a New Study Finds," Inside Climate News, April 2, 2021. https://insideclimatenews.org/news/02042021/meat-dairy-lobby-climate-action/.

87 **has also spent over $2.5 billion:** Open Secrets, Inci Sayki and Jimmy Cloutier, "Oil and Gas Industry Spent $124.4 Million on Federal Lobbying Amid Record Profits in 2022," February 22, 2023. https://www.opensecrets.org/news/2023/02/oil-and-gas-industry-spent-124-4-million-on-federal-lobbying-amid-record-profits-in-2022/.

88 **the industry spent over $132 million:** Open Secrets, "Industry Profile: Oil & Gas." https://www.opensecrets.org/federal-lobbying/industries/summary?cycle=2023&id=E01.

89 **bribery of public officials in Illinois:** Julia Gheorghiu, "ComEd Admits to Bribery Charge in Illinois, Agrees to Pay $200M Fine," Utility Dive, July 20, 2020. https://www.utilitydive.com/news/comed-admits-to-bribery-charge-in-illinois-agrees-to-pay-200m-fine/581895/.

90 **$19 million on lobbying:** Open Secrets, "Meta." https://www.opensecrets.org/orgs/facebook-inc/summary?id=D000033563.

91 **Google spent $14 million:** Open Secrets, "Alphabet Inc." https://www.opensecrets.org/orgs/alphabet-inc/summary?id=D000067823.

92 **spent roughly $10 million:** Open Secrets, "Microsoft Corp." https://www.opensecrets.org/orgs/microsoft-corp/summary?id=D000000115.

93 **on lobbying and campaign contributions:** Lauren Feiner, "Apple Ramped Up Lobbying Spending in 2022, Outpacing Tech Peers." CNBC, January 23, 2023. https://www.cnbc.com/2023/01/23/apple-ramped-up-lobbying-spending-in-2022-outpacing-tech-peers.html#:~:text=Apple%20grew%20its%20lobbying%20spend,the%20federal%20government%20in%202022.

94 **political action committee or lobbyist:** Jane Chung, "Big Tech, Big Cash: Washington's New Power Players," Public Citizen, March 24, 2021. https://www.citizen.org/article/big-tech-lobbying-update/.

95 **third of all global greenhouse emissions:** M. Crippa et al., "Food Systems Are Responsible for a Third of Global Anthropogenic GHG Emissions," *Nature Food*, March 8, 2021. https://www.nature.com/articles/s43016-021-00225-9.

96 **In a speech in 2023:** United Nations, "Secretary-General's Remarks at the World Economic Forum," January 18, 2023. https://www.un.org/sg/en/content/sg/statement/2023-01-18/secretary-generals-remarks-the-world-economic-forum.

97 **Meta has allowed violent extremist groups:** Josh Lipowsky and Gretchen Peters, "We've Tracked Extremist Content on Facebook for Years: It Doesn't Get Removed for Long," Morning Consult, November 9, 2020. https://morningconsult.com/opinions/weve-tracked-extremist-content-on-facebook-for-years-it-doesnt-get-removed-for-long/.

98 **used to incite genocides in Myanmar:** Paul Mozur, "A Genocide Incited on Facebook, with Posts from Myanmar's Military," *New York Times*, October 15, 2018. https://www.nytimes.com/2018/10/15/technology/myanmar-facebook-genocide.html.

99 **Facebook's algorithm recommended it to them:** Jeff Horwitz and Deepa Seetharaman, "Facebook Executives Shut Down Efforts to Make the Site Less Divisive," *Wall Street Journal*, May 26, 2020. https://www.wsj.com/articles/facebook-knows-it-encourages-division-top-executives-nixed-solutions-11590507499.

100 **reports of child sexual abuse material:** Katie McQue, "Child Safety Groups and Prosecutors Criticize Encryption of Facebook and Messenger," *The Guardian*, December 8, 2023. https://www.theguardian.com/technology/2023/dec/08/facebook-messenger-encryption-child-sexual-abuse.

101 **report by the New York Times:** Kate Conger, "Terrorists Are Paying for Check Marks on X, Report Says," *New York Times*, February 14, 2024. https://www.nytimes.com/2024/02/14/technology/terrorists-check-marks-x-report.html.

102 **users for hours on January 6, 2021:** Jon Levine, "Twitter Allows 'Hang Mike Pence' to Trend Hours After Trump Ban," *New York Post*, January 9, 2021. https://nypost.com/2021/01/09/twitter-allows-hang-mike-pence-to-trend-hours-after-trump-ban/.

103 **class-action lawsuit over chicken price-fixing:** CohenMilstein, "Tyson Has $99 Million Deal with Consumers in Chicken Cartel Case," March 2, 2021. https://www.cohenmilstein.com/update/"tyson-has-99-million-deal-consumers-chicken-cartel-case"-bloomberg-law.

104 **limit competition and manipulate prices:** Kaufman, "Is It Time to Break Up Big Ag?"

105 "DFA Scores Record Net Income, Despite Dip in U.S. Milk Prices," *Ingram's*. https://ingrams.com/article/dfa-nets-record-income-despite-dip-in-prices/.

106 **spill that ravaged the Gulf of Mexico:** Associated Press and Tim Stelloh, "Judge Approves $20 Billion Settlement in BP Oil Spill," NBC News, April 4, 2016. https://www.nbcnews.com/business/business-news/judge-approves-20-billion-settlement-bp-oil-spill-n550456#.

107 **BP's net profits were over $15 billion:** Statista Research Department, "Net Income/Loss of BP 2005 to 2023," April 4, 2024. https://www.statista.com/statistics/268399/profit-of-bp-since-2003/#:~:text=BP%20reported%20a%20net%20profit,demand%20due%20to%20the%20pandemic.

108 **catastrophic 2017 and 2018 wildfires:** "PG&E Hit with $2 Billion Penalty for California Wildfires," *The Independent*, May 15, 2020. https://www.independentnews.com/news/pg-e-hit-with-2-billion-penalty-for-california-wildfires/article_84a0ce38-96dd-11ea-a991-537c4048a5f5.html.

109 **saw net profits of over $2 billion:** Macrotrends, "Pacific Gas & Electric Net Income 2010–2023 | PCG." https://www.macrotrends.net/stocks/charts/PCG/pacific-gas-electric/net-income.

110 **wake of the Cambridge Analytica scandal:** Cecilia Kang, "F.T.C. Approves Facebook Fine of About $5 Billion," *New York Times*, July 12, 2019. https://www.nytimes.com/2019/07/12/technology/facebook-ftc-fine.html#:~:text=The%20Federal%20Trade%20Commission%20has,the%20country's%20most%20powerful%20technology.

111 **imposed on Meta was approximately $7.1 trillion:** Craig Timberg and Tony Romm, "How Big Could Facebook's Fine Theoretically Get? Here's a Hint: There Are Four Commas, and Counting," *Washington Post*, April 9, 2018. https://www.washingtonpost.com/news/the-switch/wp/2018/04/09/how-big-could-facebooks-fine-theoretically-get-heres-a-hint-there-are-four-commas-and-counting/?noredirect=on.

112 **$5.8 trillion US money in circulation, total:** Rebekah Carter, "How Much Money Is in Circulation in 2023?" MoneyTransfers, October 6, 2023. https://moneytransfers.com/news/2022/06/12/how-much-money-is-in-circulation.

Chapter 3: Promising Movements Afoot

1 **"higher quality product onto our tables":** US Energy, "Gabe Brown Discusses How Regenerative Agriculture Is a Solution to Global Challenges," January 13, 2021. https://www.youtube.com/watch?v=TLwsn8snsMc.

2 **carbon dioxide from the Earth's atmosphere:** EarthDay Organization, "Food Doesn't Grow on Its Own." Accessed June 2, 2024. https://www.earthday.org/campaign/regenerative-agriculture/.
3 **solar array was introduced in 1956:** Everlight Solar, "A Brief History of Solar Power. How Did It Come to Be?" November 29, 2021. https://everlightsolar.com/a-brief-history-of-solar-power-how-did-it-come-to-be/.
4 **as little as $0.50 per watt:** Luke Richardson, "Solar History: Timeline & Invention of Solar Panels," EnergySage, April 26, 2023. https://www.energysage.com/about-clean-energy/solar/the-history-and-invention-of-solar-panel-technology/.
5 **(and overtake coal) by 2025:** IEA, "Electricity Market Report 2023," March 2023. https://iea.blob.core.windows.net/assets/255e9cba-da84-4681-8c1f-458ca1a3d9ca/ElectricityMarketReport2023.pdf.
6 **recalling all EV1's and crushing them:** BBC, "EV1: How an Electric Car Dream Was Crushed," January 28, 2022. https://www.bbc.com/news/av/stories-60160474.
7 **18% of global car sales in 2023:** Marcus Lu, "Ranked: Electric Vehicle Sales by Model in 2023," *Visual Capitalist*, December 1, 2023. https://www.visualcapitalist.com/electric-vehicle-sales-by-model-2023/.
8 **sales of gas and diesel cars globally by 2040:** Caleb Miller, "Six Major Automakers Agree to End Gas Car Sales Globally by 2040," *Car and Driver*, November 11, 2021. https://www.caranddriver.com/news/a38213848/automakers-pledge-end-gas-sales-2040/.
9 **Moore's:** Carla Tardi, "What Is Moore's Law and Is It Still True?" *Investopedia*, April 2, 2024. https://www.investopedia.com/terms/m/mooreslaw.asp.
10 **Koomey's Laws:** "Koomey's Law," *Semiconductor Engineering*. Accessed June 2, 2024. https://semiengineering.com/knowledge_centers/standards-laws/laws/koomeys-law/.

Chapter 4: Web3 Is Here: What the Heck Is It?

1 **"you don't need to trust a single organization":** CNBC International, "What Is Web3, and Is It the Future of the Internet?" June 6, 2022. https://www.youtube.com/watch?v=eAMCcnxXLhM.
2 **for what would become the World Wide Web:** CERN, "The Birth of the Web." Accessed June 2, 2024. https://home.web.cern.ch/science/computing/birth-web#:~:text=Tim%20Berners%2DLee%2C%20a%20British,and%20institutes%20around%20the%20world.

3 **in the 1990s:** SuperGroups.com, "The First Community Portal." Accessed June 2, 2024. http://supergroups2001.com/SuperGroupsPPT_6-16-11.pdf.

4 **"A Peer-to-Peer Electronic Cash System":** Michael Adams, "Who Is Satoshi Nakamoto?" *Forbes*, March 18, 2023. https://www.forbes.com/advisor/investing/cryptocurrency/who-is-satoshi-nakamoto/.

5 **real estate, digital art, games, and more:** Adam Levy, "15 Applications for Blockchain Technology," The Motley Fool, November 7, 2023. https://www.fool.com/investing/stock-market/market-sectors/financials/blockchain-stocks/blockchain-applications/.

6 **coined the term "Web3" in 2014:** Arjun Kharpal, "What Is 'Web3'? Here's the Vision for the Future of the Internet from the Man Who Coined the Phrase," CNBC, April 19, 2022. https://www.cnbc.com/2022/04/20/what-is-web3-gavin-wood-who-invented-the-word-gives-his-vision.html.

7 **the Web3 world:** Dock.io, "Dock Launches Web3 ID: Secure Web3 Identity for People and Organizations," July 15, 2022. https://blog.dock.io/web3-identity1/.

8 **plan never materialized:** Russell Brandom, "Google, Facebook, Microsoft, and Twitter Partner for Ambitious New Data Project," *The Verge*, July 20, 2018. https://www.theverge.com/2018/7/20/17589246/data-transfer-project-google-facebook-microsoft-twitter.

9 **"data from one service to another in Web3":** Reed McGinley-Stempel, "Web3 and the Future of Data Portability: Rethinking User Experiences and Incentives on the Internet," HelpNet Security, March 30, 2022. https://www.helpnetsecurity.com/2022/03/30/web3-applications/.

10 **"computers rather than on a single server":** Moralis, "Moralis Apis Powers Crypto and Blockchain Applications for Millions of End Users Worldwide." Accessed June 2, 2024. https://moralis.io/?utm_source=blog&utm_medium=post&utm_campaign=Web3%2520Storage%2520%25E2%2580%2593%2520How%2520Web3%2520Data%2520Storage%2520Works.

11 **cryptocurrency rewards for your contributions:** Johnny Lyu, "Web3 Social Networks Can't Rely on Money Incentives Alone," Be(In) Crypto, March 14, 2022. https://beincrypto.com/web3-social-networks-cant-rely-on-money-incentives-alone/.

12 **"there's an opportunity cost for trolling":** Jessica Klein, "How This Obscure, Blockchain-Based Site Built a Playground for QAnon to Run Rampant On," DailyDot, October 28, 2021. https://www.dailydot.com/debug/qanon-steemit/.

Chapter 5: A Healthy Dose of Web3 Reality

1 **"I don't have the control":** Adi Robertson, "How the Biggest Decentralized Social Network Is Dealing with Its Nazi Problem," The Verge, July 13, 2019. https://www.theverge.com/2019/7/12/20691957/mastodon-decentralized-social-network-gab-migration-fediverse-app-blocking.

2 **the text string persists forever:** McKinsey & Company, "What Is Blockchain?" December 5, 2022. https://www.mckinsey.com/featured-insights/mckinsey-explainers/what-is-blockchain#.

3 **disintegration of that space:** Vili Lehdonvirta, "Cloud Empires: How Digital Platforms Are Overtaking the State and How We Can Regain Control," Taylor and Francis Online, March 23, 2023. https://doi.org/10.1080/1369118X.2023.2193248.

4 **an average of 40% of the user base:** Marcel Pechman, "Analyst Says 40% of Users in Most Web3 Games Are Bots — Here's How to Avoid Being Fooled," Cointelegraph, August 31, 2022. https://cointelegraph.com/news/analyst-says-40-of-users-in-most-web3-games-are-bots-here-s-how-to-avoid-being-fooled.

5 **virtual world on the Ethereum blockchain:** Liam "Akida" Wright, "Study Discovers 40% of Web3 Gaming Accounts Are Bots Using on-Chain Data," CryptoSlate, August 31, 2022. https://cryptoslate.com/study-discovers-40-of-web3-gaming-accounts-are-bots-using-on-chain-data/.

6 **"limited-edition digital assets at a low price":** Ruholamin Haqshanas, "Bots Represent 40% of the Average Web3 Platform Users: Report, *Milk Road*, August 31, 2022. https://milkroad.com/news/bots-represent-40-of-the-average-web3-platform-users-report/.

7 **Chainalysis:** Reuters, "Crypto Hackers Stole Around $1.7 Bln in 2023—Report," January 25, 2024. https://www.reuters.com/technology/cybersecurity/crypto-hackers-stole-around-17-bln-2023-report-2024-01-24/#:~:text=Jan%2024%20(Reuters)%20%2D%20Hackers,Chainalysis%20report%20showed%20on%20Wednesday.

8 **"move to the top of a chain of comments":** Brady Dale, "Block.One Is Launching a Social Media Platform on the Eos Blockchain," CoinDesk, June 2, 2019. https://www.coindesk.com/markets/2019/06/01/blockone-is-launching-a-social-media-platform-on-the-eos-blockchain/#.

9 **entities seeking to disrupt civil discourse:** CORDIS, "The Secret Robot Armies Fighting to Undermine Democracy," September 28, 2021. https://cordis.europa.eu/article/id/430251-the-secret-robot-armies-fighting-to-undermine-democracy.

10 **you did not choose to follow:** Fedi.Tips, "Where Are the Trending Posts and Hashtags on Mastodon?" https://fedi.tips/where-are-the-trending-posts-and-hashtags-on-mastodon/.
11 **Odysee's privacy policy:** Odysee, "Privacy Policy," October 8, 2021. https://odysee.com/$/privacypolicy.
12 **"down all the way to their sources":** Jamila Grier, "Web3 Businesses Have Major Data Privacy Challenges Ahead," *Forbes*, February 22, 2023. https://www.forbes.com/sites/forbestechcouncil/2023/02/22/web3-businesses-have-major-data-privacy-challenges-ahead/?sh=24c776f31485.
13 **"Bitcoin anonymity is just a big myth":** Madonna Prathap, "Bitcoin Does Not Make Payments Anonymous—Just Really Hard to Trace," *Business Insider India*, December 24 2021. https://www.businessinsider.in/investment/news/bitcoin-does-not-make-payments-anonymous-just-really-hard-to-trace/articleshow/85068905.cms.
14 **"always get to the bottom of it":** Julian Dossett, "Are Cryptocurrency Transactions Actually Anonymous?" CNET, June 7, 2022. https://www.cnet.com/personal-finance/crypto/are-cryptocurrency-transactions-actually-anonymous/.
15 **"are more traceable than cash":** Shalini Nagarajan, "Bitcoin Anonymity Is Just a Big Myth—And Using It to Launder Dirty Money Is Stupid, a Crypto ATM Chief Says," *Business Insider*, June 13, 2021. https://markets.businessinsider.com/currencies/news/bitcoin-anonymous-untraceable-myths-stupid-dirty-money-laundering-crypto-chief-2021-6-1030517840?miRedirects=1.
16 **determine someone's real identity:** Mark Gibbs, "MIT Researchers Show You Can Be Identified by a Just Few [*sic*] Data Points," NetworkWorld, January 30, 2015. https://www.networkworld.com/article/935140/mit-researchers-show-you-can-be-identified-by-a-just-few-data-points.html.
17 **"stores have to be fast, cheap and private":** Ryan Browne, "Web Inventor Tim Berners-Lee Wants Us to 'Ignore' Web3: 'Web3 Is Not the Web at All,'" CNBC, November 4, 2022. https://www.cnbc.com/2022/11/04/web-inventor-tim-berners-lee-wants-us-to-ignore-web3.html#:~:text=Tim%20Berners%2DLee%2C%20the%20computer,Summit%20tech%20conference%20in%20Lisbon.
18 **social network, emerging in 2016:** SoFi, "What Is Steemit (STEEM)?" April 12, 2022. https://www.sofi.com/what-is-steemit/.
19 **between 2022 and 2023:** Chris Metinko, "Web3 Funding Cratered in 2023," Crunchbase News, January 22, 2024. https://news.crunchbase.com/web3/funding-cratered-sbf-ai-crypto-bitcoin-eoy-2023/.

20 **1 percent of token holders:** Ephrat Livni, "Tales from Crypto: A Billionaire Meme Feud Threatens Industry Unity," *New York Times*, January 18, 2022. https://www.nytimes.com/2022/01/18/business/dealbook/web3-venture-capital-andreessen.html.

21 **"90% of its revenue from transaction fees":** Jamie Crawley, "Binance Generates 90% of Revenue from Transaction Fees, Changpeng Zhao Says," CoinDesk, December 7, 2022. https://www.coindesk.com/business/2022/12/07/binance-generates-90-of-revenue-from-transaction-fees-changpeng-zhao-says/.

22 **banned cryptocurrency entirely:** Kieth Rean Garcia, "List of Countries Banning Cryptocurrency," CryptoTicker.io, May 3, 2023. https://cryptoticker.io/en/banned-cryptocurrency-countries-2023/.

23 **"run very well on PowerPoint":** Peter Coy, "Yes, the Internet Is Broken, but What Does a Fix Look Like?" *New York Times*, March 13, 2024, https://www.nytimes.com/2024/03/13/opinion/internet-web-3-tech.html.

24 **"without the need for intermediaries":** UNICEF, "Blockchain: Exploring Blockchain Applications to Accelerate Impact." https://www.unicef.org/innovation/blockchain.

25 **fair trade coffee:** Andry Alamsyah et al., "Blockchain Traceability Model in the Coffee Industry," Journal of Open Innovation: Technology, Market, and Complexity, September 2023. https://www.sciencedirect.com/science/article/pii/S2199853123001105.

26 **palm oil:** Anna Hofer, "Four Examples of Blockchain in Supply Chain Management," Softeq, March 1, 2023, https://www.softeq.com/blog/four-blockchain-supply-chain-examples.

27 **conflict-free diamonds:** Terence Zimwara, "Blockchain Technology Can Guarantee to Consumers 'That Their Diamonds Have Been Ethically Sourced'—Botswana President," Bitcoin.com, October 31, 2023, https://news.bitcoin.com/blockchain-technology-can-guarantee-to-consumers-that-their-diamonds-have-been-ethically-sourced-botswana-president/.

28 **sourcing of their purchased items:** Sam Daley, "Blockchain in Healthcare: 18 Examples to Know," BuiltIn, February 16, 2023. https://builtin.com/blockchain/blockchain-healthcare-applications-companies.

29 **"this is regression, not progress":** Stephen Diehl, "Web3 Is Bullshit." https://www.stephendiehl.com/blog/web3-bullshit.html.

Chapter 6: Meet Your New Boss, Same as the Old Boss

1 **Jack Dorsey:** Alex Hern, "Jack Dorsey Quits Bluesky Board and Urges Users to Stay on Elon Musk's X," *The Guardian*, May 7, 2024. https://www.theguardian.com/technology/article/2024/may/07/jack-dorsey-quits-bluesky-board-urges-users-stay-elon-musk-x-twitter.

2 **Horowitz is backing several Web3 social sites:** Ezra Regeurra, "Andreessen Horowitz Exec Sees 'Promising Paths' for Web3 Social Platforms—EthCC," Cointelegraph, July 20, 2023. https://cointelegraph.com/news/web-3-andreessen-horowitz-exec-sees-promising-paths-for-web3-social-platforms-eth-cc-6.

3 **"effectively more of a plutocracy than democracy":** Justin Mart and Ryan Yi, "Around the Block #13: On the Value and Risks of Governance Tokens." Coinbase. Accessed June 2, 2024. https://www.coinbase.com/learn/market-updates/around-the-block-issue-13.

4 **the fray with its Twitter/X clone, Threads:** Maria Diaz, "Threads Becomes Fastest-Growing App Ever, with 100 Million Users in Under a Week," ZDNet, July 10, 2023. https://www.zdnet.com/article/threads-hit-100-million-users-in-under-a-week-breaking-chatgpts-record/.

5 **financial info, health and fitness, and more:** Jackie Ruryk, "Are You on Threads Yet? Here's What You're Giving Away," CBC, July 7, 2023. https://www.cbc.ca/news/world/threads-privacy-terms-of-service-1.6900104.

6 **the main squeeze for nonfungible token metadata:** Russell Brandom, "How One Company Took over the NFT Trade," The Verge, February 2, 2022. https://www.theverge.com/2022/2/2/22914081/open-sea-nft-marketplace-web3-fundraising-finzer-a16z.

7 **"proponents decry about Web 2.0.":** Peter Coy, "Yes, the Internet Is Broken, but What Does a Fix Look Like?" *New York Times*, March 13, 2024. https://www.nytimes.com/2024/03/13/opinion/internet-web-3-tech.html.

Chapter 7: Is There a Better Way?

1 **Americans had confidence in big business:** Lydia Saad, "Confidence in Big Business, Big Tech Wanes Among Republicans," Gallup, July 19, 2021. https://news.gallup.com/poll/352484/confidence-big-business-big-tech-wanes-among-republicans.aspx.

2 **three Americans had a positive view of Big Tech:** Megan Brenan, "Views of Big Tech Worsen; Public Wants More Regulation," Gallup, February 18, 2021. https://news.gallup.com/poll/329666/views-big-tech-worsen-public-wants-regulation.aspx.

3 **Whole Foods cofounder John Mackey:** John Mackey and Rajendra Sisodia, *Conscious Capitalism, with a New Preface by the Authors: Liberating the Heroic Spirit of Business*. Harvard Business Review Press, 2014. https://www.amazon.com/Conscious-Capitalism-New-Preface-Authors/dp/1625271751.

4 **"people out of poverty and creates prosperity":** Conscious Capitalism, "We Believe That Business Is Good." Accessed June 2, 2024. https://consciouscapitalismcmd.org/about-the-movement/.

5 **perform 10 times better than their peers:** Bronwyn Fryer, "The Rise of Compassionate Management (Finally)," *Harvard Business Review*, September 18, 2013. https://hbr.org/2013/09/the-rise-of-compassionate-management-finally#:~:text=It%27s%20not%20just%20altruism%3A%20as,are%20finally%20making%20a%20dent.

6 **"markets and makes [us] more money":** Jeff Beer, "How Patagonia Grows Every Time It Amplifies Its Social Mission," *Fast Company*, February 21, 2018. https://www.fastcompany.com/40525452/how-patagonia-grows-every-time-it-amplifies-its-social-mission#:~:text="Doing%20good%20work%20for%20the,priority%20in%20the%20corporate%20world.

7 **over 70 million users:** Bojan Jovanovic, "DuckDuckgo Statistics: Google's Avian Adversary in Numbers," DataProt, May 6, 2023. https://dataprot.net/statistics/duckduckgo-statistics/.

8 **(that's trillion with a T!):** Ycharts, "Apple Market Cap." Accessed June 2, 2024. https://ycharts.com/companies/AAPL/market_cap.

9 **portion of its profits with employees:** Corinne Lestch, "Eileen Fisher Likes to Share Her Company with Employees Who 'Do All the Work,'" The Story Exchange, January 9, 2020. https://thestoryexchange.org/eileen-fisher-employees-have-strong-stake-fashion-brand/.

10 **environmental nonprofit organizations:** Patagonia, "1% for the Planet." https://www.patagonia.com/one-percent-for-the-planet.html#:~:text=Since%201985%2C%20Patagonia%20has%20pledged,difference%20in%20their%20local%20communities.

11 **transition to regenerative farming:** Esha Chhabra, "Chocolate Brand Looks to Carbon Insetting to Transition Farmers to Regenerative Agriculture," *Forbes*, January 31, 2021. https://www.forbes.com/sites/eshachhabra/2021/01/31/chocolate-brand-looks-to-carbon-insetting-to-transition-farmers-to-regenerative-agriculture/?sh=31eedc157e7b.

12 **his Solid protocol initiative:** Arjun Kharpal, "The Inventor of the Web Thinks Everyone Will Have Their Own Personal A.I. Assistants Like ChatGPT," CNBC, February 16, 2023. https://www.cnbc.com/2023/02/17/tim-berners-lee-thinks-we-will-have-our-own-ai-assistants-like-chatgpt.html.

13 **Pods for all its citizens:** Peter Coy, "Yes, The Internet Is Broken, But What Does a Fix Look Like?" *New York Times*, March 13, 2024. https://www.nytimes.com/2024/03/13/opinion/internet-web-3-tech.html.

14 **able to acquire them:** Robert Peck, "Reddit Is Letting Power Users in on Its IPO. Not Everyone's Buying," WIRED, February 23, 2024. https://www.wired.com/story/reddit-power-users-ipo/.

Chapter 8: Empowering You: Social Media User, Creator, Star

1 **CNBC reported in 2023:** Jennifer Liu, "More Than Half of Gen Zers Think They 'Can Easily Make a Career in Influencing,' Says Branding Expert," CNBC, September 20, 2023. https://www.cnbc.com/2023/09/20/more-than-half-of-gen-zers-think-they-can-easily-make-a-career-in-influencing.html#:~:text=Most%20young%20people%20today%20see,Consult%2C%20a%20decision%20intelligence%20company.

2 **creator economy just since 2020:** Jasmine Enberg, "5 Things to Know About the Creator Economy in 2023," *Insider Intelligence*, February 23, 2023. https://www.insiderintelligence.com/content/5-things-know-about-creator-economy-2023.

3 **$80 million per year:** Eleanor Pringle, "MrBeast Says It's 'Painful' Watching Wannabe YouTube Influencers Quit School and Jobs for a Pipe Dream: 'For Every Person Like Me That Makes It, Thousands Don't,'' *Fortune*, March 15, 2024. https://fortune.com/2024/03/15/youtube-biggest-star-mrbeast-jimmy-donaldson-painful-watch-would-be-influencers-quit-school-jobs-pipe-dream/.

4 **look at "That Chick Angel":** Zach Seemayer, "One Margarita' Rapper on Song's Viral Success and Landing Cindy Crawford for the Music Video (Exclusive)," *ET Online*, August 1, 2023. https://www.etonline.com/one-margarita-rapper-on-songs-viral-success-and-landing-cindy-crawford-for-the-music-video.

5 **thousands don't:** Pringle, "MrBeast Says It's 'Painful' Watching Wannabe YouTube Influencers."

6 **2023 report by Linktree:** Amanda Silberling, "Only 12% of Full-Time Creators Make Over $50k a Year, Says Linktree," *TechCrunch*, April 20,

2022. https://techcrunch.com/2022/04/20/linktree-creator-economy-report-research/.

7 **5% of its page followers:** Neil Patel, "20 Secrets to Boost Your Facebook Organic Reach." Accessed June 2, 2024. https://neilpatel.com/blog/13-secrets-thatll-boost-your-facebook-organic-reach/.

8 **"data interoperability":** Intertrust, "What Is Data Interoperability?" January 4, 2023. https://www.intertrust.com/blog/what-is-data-interoperability/.

9 **Restoration Network and ActivityPub:** David Pierce, "Can ActivityPub Save the Internet?" The Verge, April 20, 2023. https://www.theverge.com/2023/4/20/23689570/activitypub-protocol-standard-social-network.

Chapter 9: Do I Need a Social Media Wallet: Is There Money in It?

1 **their prices fluctuate wildly:** Erik Anderson, "Investors Fear Volatility and Risk, Particularly with Crypto. Here's Why They Shouldn't," CoinDesk, October 4, 2023. https://www.coindesk.com/markets/2023/10/04/volatility-and-risk-get-a-bad-rap-in-investment-portfolios-particularly-when-it-comes-to-digital-assets-heres-why-they-shouldnt/.

2 **it would be ending Reddit Coins:** Jay Peters, "Reddit Is Getting Rid of Its Gold Awards System," The Verge, July 14, 2023, https://www.theverge.com/2023/7/13/23794403/reddit-gold-awards-coins-sunset.

Chapter 10: How Do We Overcome C-Suite Tyranny?

1 **offer feedback on product development:** Microsoft, "Microsoft Community Advisory Board." Accessed June 2, 2024. https://adoption.microsoft.com/en-us/community/advisory-board/.

2 **hosts from various countries:** Airbnb, "Airbnb Announces the 2023 Host Advisory Board," February 1, 2023. https://news.airbnb.com/airbnb-announces-2023-host-advisory-board/.

3 **compliance with regenerative principles:** Rodale Institute, "The Global Leader of Regenerative Organic Agriculture." Accessed June 2, 2024. https://rodaleinstitute.org/wp-content/uploads/Rodale-Press-Kit-2020_Digital.pdf.

4 **renewable energy and carbon offset products:** Green-E Org, "Welcome to Green-e Certification." Accessed June 2, 2024. https://www.green-e.org.

5 **so much for helpful "oversight"!:** Elizabeth Dwoskin, "Facebook Tried to Outsource Its Decision About Trump. The Oversight Board Said Not So Fast," *Washington Post*, May 6, 2021. https://www.washingtonpost.com/technology/2021/05/06/facebook-oversight-board-punted-back/.

6 **deepfakes incoherent:** Katie Paul, "Meta Oversight Board Calls Company's Deepfake Rule 'Incoherent,'" Reuters, February 5, 2024. https://www.reuters.com/technology/cybersecurity/meta-oversight-board-reviewing-biden-video-says-deepfake-rule-incoherent-2024-02-05/.

Chapter 11: In the Crosshairs: Privacy and Anonymity

1 **naked photos of themselves:** "Facebook: Send Us Your Naked Photos to Stop Revenge Porn," *CBS News*, May 24, 2018. https://www.cbsnews.com/news/facebook-revenge-porn-naked-photos-pilot-progam/.

2 **massive data leaks:** Emma Bowman, "After Data Breach Exposes 530 Million, Facebook Says It Will Not Notify Users," NPR, April 9, 2021. https://www.npr.org/2021/04/09/986005820/after-data-breach-exposes-530-million-facebook-says-it-will-not-notify-users.

3 **"your personal information is collected and used":** IAAP, "What Is Privacy." Accessed June 2, 2024. https://iapp.org/about/what-is-privacy/.

4 **Tim Berners-Lee, inventor of the World Wide Web:** Sam Thielman, "Tim Berners-Lee: Selling Private Citizens' Browsing Data Is 'Disgusting,'" *The Guardian*, April 4, 2017. https://www.theguardian.com/technology/2017/apr/04/tim-berners-lee-online-privacy-interview-turing-award.

5 **Tim Cook, CEO of Apple:** Marvin the Robot, "10 Best Quotes from Tim Cook's Recent Speech on Privacy and Security," *Kaspersky*, June 3, 2015. https://usa.kaspersky.com/blog/tim-cook-speaks-about-privacy-security/5391/.

6 **Cavoukian, former Privacy Commissioner of Ontario:** *Socrates*, "Privacy Commissioner Dr. Ann Cavoukian: We Have to Protect Privacy Globally or We Protect It Nowhere!" *Singularity Weblog*, July 26, 2013. https://www.singularityweblog.com/ann-cavoukian-privacy-by-design/.

7 **Snowden, former NSA contractor and whistleblower:** Paul Schrodt, "Edward Snowden Just Made an Impassioned Argument for Why Privacy Is the Most Important Right," *Business Insider*, September 16, 2016. https://www.businessinsider.com/edward-snowden-privacy-argument-2016-9.

8 **Eric Schmidt, former CEO of Google:** Richard Esguerra, "Google CEO Eric Schmidt Dismisses the Importance of Privacy," Electronic Frontier Foundation, December 10, 2009. https://www.eff.org/deeplinks/2009/12/google-ceo-eric-schmidt-dismisses-privacy.

9 **Mark Zuckerberg, CEO of Meta:** Marshall Kirkpatrick, "Facebook's Zuckerberg Says the Age of Privacy Is Over," *New York Times*, January 10, 2010. https://archive.nytimes.com/www.nytimes.com/external/readwriteweb/2010/01/10/10readwriteweb-facebooks-zuckerberg-says-the-age-of-privac-82963.html?source=post%255Fpage.

10 **McNealy, former CEO of Sun Microsystems:** Polly Sprenger, "Sun On Privacy: 'Get Over It,'" WIRED, January 26, 1999. https://www.wired.com/1999/01/sun-on-privacy-get-over-it/.

11 **Cashmore, founder of Mashable:** Pete Cashmore, "Privacy Is Dead, and Social Media Hold Smoking Gun," CNN, October 28, 2009. https://edition.cnn.com/2009/OPINION/10/28/cashmore.online.privacy/.

12 **"to preserve and enlarge freedom":** Oxford Reference, "John Locke 1632–1704 English Philosopher." Accessed June 2, 2024. https://www.oxfordreference.com/display/10.1093/acref/9780191826719.001.0001/q-oro-ed4-00006743.

13 **"surrender the rights of humanity":** Fordham University, "Internet Modern History Sourcebook: Jean-Jacques Rousseau: The Social Contract, 1762, Extended Excerpts." Accessed June 2, 2024. https://sourcebooks.fordham.edu/mod/rousseau-contract2.asp.

14 **encompass the right to privacy:** Annenberg Classroom, "Chapter 22: The Right of Privacy." Accessed June 2, 2024. https://www.annenbergclassroom.org/resource/our-rights/rights-chapter-22-right-privacy/.

15 **("the right to be let alone"):** Samuel D. Warren and Louis D. Brandeis, "The Right to Privacy," *Harvard Law Review* 4, No. 5 (December 15, 1890). https://www.jstor.org/stable/1321160.

16 **"the privacy of the individual":** University of Michigan, "History of Privacy Timeline." Accessed June 2, 2024. https://safecomputing.umich.edu/privacy/history-of-privacy-timeline.

17 **government surveillance and censorship of its citizens:** Richard Esguerra, "Google CEO Eric Schmidt Dismisses the Importance of Privacy," Electronic Frontier Foundation, December 10, 2009. eff.org/deeplinks/2009/12/google-ceo-eric-schmidt-dismisses-privacy.

18 **"such interference or attacks":** United Nations, "Universal Declaration of Human Rights."Accessed June 2, 2024. https://www.un.org/en/about-us/universal-declaration-of-human-rights.

19 ***Griswold v. Connecticut* in 1965:** National Constitution Center, "Griswold v. Connecticut (1965)." https://constitutioncenter.org/the-constitution/supreme-court-case-library/griswold-v-connecticut.

20 **people, not just places:** Oyez, "Katz v. United States." Accessed June 2, 2024. https://www.oyez.org/cases/1967/35.

21 **outspoken privacy advocate:** Pieter Verdegem, "Tim Berners-Lee's Plan to Save the Internet: Give Us Back Control of Our Data," *The Conversation*, February 5, 2021. https://theconversation.com/tim-berners-lees-plan-to-save-the-internet-give-us-back-control-of-our-data-154130.

22 **initiatives like Pods and Solid:** Solid, "What Is Solid?" Accessed June 2, 2024. https://solid.mit.edu.

23 **providers and insurance companies:** Centers for Disease Control and Prevention, "Health Insurance Portability And Accountability Act of 1996 (HIPAA)." Accessed June 2, 2024. https://www.cdc.gov/phlp/publications/topic/hipaa.html.

24 **(COPPA), passed in 1998:** Federal Trade Commission, "Children's Online Privacy Protection Rule ('COPPA')." Accessed June 2, 2024. https://www.ftc.gov/legal-library/browse/rules/childrens-online-privacy-protection-rule-coppa.

25 **(GDPR), which went into effect in 2018:** European Union, "General Data Protection Regulation (GDPR)." Accessed June 2, 2024. https://gdpr-info.eu.

26 **unveiled in 2024:** Mark Weinstein, "Opinion: Americans Might Finally Get a Real Privacy Law to Fight Big Tech Intrusions," *Los Angeles Times*, April 29, 2024. https://www.latimes.com/opinion/story/2024-04-29/americans-privacy-law-big-tech.

27 **enshrined by the First Amendment:** Electronic Frontier Foundation, "Anonymity." https://www.eff.org/issues/anonymity.

28 **"the hand of an intolerant society":** Ibid.

29 **Protection Act in the United States:** Candice Norwood, "Whistleblower Protection, Explained," *PBS Newshour*, October 2, 2019. https://www.pbs.org/newshour/politics/whistleblower-protections-explained.

30 **used by billions of people worldwide:** Mark Weinstein, "The Government Wants Access to Your Encrypted Messages—We Must Act Now to Defend Our Right to Privacy," The Hill, July 6, 2023. https://thehill.com/opinion/technology/4081907-the-government-wants-access-to-your-encrypted-messages-we-must-act-now-to-defend-our-right-to-privacy/.

31 **"client-side scanning":** Electronic Frontier Foundation, Joe Mullen, "The EARN IT Bill is Back, Seeking to Scan Our Messages and Photos,"

April 20, 2023. https://www.eff.org/deeplinks/2023/04/earn-it-bill-back-again-seeking-scan-our-messages-and-photos.

32 **exploitation online in 2023:** National Center for Missing & Exploited Children, "CyberTipline 2022 Report." Accessed June 2, 2024. https://www.missingkids.org/cybertiplinedata.

33 **95% of these reports:** Katie McQue, "Child Safety Groups and Prosecutors Criticize Encryption of Facebook and Messenger," *The Guardian*, December 8, 2023. https://www.theguardian.com/technology/2023/dec/08/facebook-messenger-encryption-child-sexual-abuse.

34 **Stanford Internet Observatory:** Riana Pfefferkorn, "The Stop CSAM Act: An Anti-Encryption Stalking Horse," Center for Internet and Society, April 29, 2023. https://cyberlaw.stanford.edu/blog/2023/04/stop-csam-act-anti-encryption-stalking-horse.

35 **a 2022 report by the New York Times:** Joe Mullin, "Google's Scans of Private Photos Led to False Accusations of Child Abuse," Electronic Frontier Foundation, August 22, 2022. https://www.eff.org/deeplinks/2022/08/googles-scans-private-photos-led-false-accusations-child-abuse.

36 **Stanford addressed a letter to the Senate:** Riana Pfefferkorn, "The EARN It Act Is Unconstitutional: Fourth Amendment," Center for Internet and Society, March 10, 2020. https://cyberlaw.stanford.edu/blog/2020/03/earn-it-act-unconstitutional-fourth-amendment.

37 ***Bernstein v. the U.S. Department of State*****:** Kevin Featherly, "Bernstein v. the U.S. Department of State Law Case," *Brittanica*. Accessed June 2, 2024. https://www.britannica.com/event/Bernstein-vs-the-US-Department-of-State.

38 **at Stanford Law School:** Pfefferkorn, "The Stop CSAM Act.

39 **platforms for such cases:** National Center for Missing & Exploited Children, CyberTipline. Accessed June 2, 2024. https://www.missingkids.org/gethelpnow/cybertipline.

Chapter 12: User ID Verification: Friend or Foe?

1 **"offensive comments rise to the top":** Christopher Wolf, "Anonymity May Have Killed Online Commenting," *New York Times*, April 18, 2016. https://www.nytimes.com/roomfordebate/2016/04/18/have-comment-sections-on-news-media-websites-failed/anonymity-may-have-killed-online-commenting.

2 **routinely wreak havoc:** Jeanna Matthews, "Bots and Trolls Control a Shocking Amount of Online Conversation," Fast Company, June 29, 2020. https://www.fastcompany.com/90521340/bots-and-trolls-control-a-shocking-amount-of-online-conversation.

3 **"national identity system":** Ashley Rindsberg, "Obama Intel Chief Wants National ID System," *HuffPost*, April 9, 2009. https://www.huffpost.com/entry/obama-intel-chief-wants-n_b_172832.
4 **8–12 use social media:** The US Surgeon General's Advisory, "Social Media and Youth Mental Health," 2023. https://www.hhs.gov/sites/default/files/sg-youth-mental-health-social-media-advisory.pdf.
5 **identity verification model:** Tom Gerken, "Tinder Introduces Passport-Scanning ID Checks for UK Users" BBC, January 31, 2024. https://www.bbc.com/news/technology-68312396.
6 **"you have had an account for at least 90 days":** X Help Center, "How to Get the Blue Checkmark on X." https://help.twitter.com/en/managing-your-account/about-x-verified-accounts.
7 *Washington Post* **in early 2023:** Geoffrey A. Fowler, "We Got Twitter 'Verified' in Minutes Posing as a Comedian and a Senator," *Washington Post*, November 11, 2022. https://www.washingtonpost.com/technology/2022/11/11/twitter-blue-checkmark/.
8 **TechCrunch in late 2023:** Sarah Perez, "Elon Musk Says X Will Charge Users 'a Small Monthly Payment' to Use Its Service," *TechCrunch*, September 18, 2023. https://techcrunch.com/2023/09/18/elon-musk-says-x-will-charge-users-a-small-monthly-payment-to-use-its-service/.
9 **government's mass surveillance practices:** Glenn Greenwald, Ewen MacAskill, and Laura Poitras, "Edward Snowden: The Whistleblower Behind the NSA Surveillance Revelations," *The Guardian*, June 11, 2013. https://www.theguardian.com/world/2013/jun/09/edward-snowden-nsa-whistleblower-surveillance.
10 **reauthorized in 2024:** "Biden Signs Bill Reauthorizing Contentious FISA Surveillance Program," *CBS News*, April 20, 2024. https://www.cbsnews.com/news/biden-signs-bill-reauthorizing-fisa-surveillance-program-section-702/.
11 **"be held accountable for what they post":** E+T Editorial Team, "Social Media Users Should Be Verified with Real ID, IT Professionals Say," *Engineering and Technology*, April 29, 2021. https://eandt.theiet.org/2021/04/29/social-media-users-should-be-verified-real-id-it-professionals-say.

Chapter 13: Saving Our Kids from the Abyss

1 **over 97% of teens:** Monica Anderson and Jingjing Jiang, "Teens' Social Media Habits and Experiences," Pew Research Center, November 28, 2018. https://www.pewresearch.org/internet/2018/11/28/teens-social-media-habits-and-experiences/.

2 **social apps each day:** Brad Adgate, "Gallup: Teens Spend More Time on Social Media Than on Homework," *Forbes*, October 18, 2023. https://www.forbes.com/sites/bradadgate/2023/10/18/gallup-teens-spend-more-time-on-social-media-than-on-homework/?sh=7cde1d5a3dcb.

3 **over the past decade:** Pew Research Center Fact Sheet, "Teens and Internet, Device Access Fact Sheet," January 5, 2024. https://www.pewresearch.org/internet/fact-sheet/teens-and-internet-device-access-fact-sheet/.

4 **benefits to many teens:** Kirsten Weir, "Social Media Brings Benefits and Risks to Teens," *Monitor on Psychology*, September 1, 2023. https://www.apa.org/monitor/2023/09/protecting-teens-on-social-media.

5 **survey by Pew Research:** Monica Anderson, "Connection, Creativity and Drama: Teen Life on Social Media in 2022," Pew Research, November 16, 2022. https://www.pewresearch.org/internet/2022/11/16/connection-creativity-and-drama-teen-life-on-social-media-in-2022/.

6 **depression or anxiety:** US Department of Health and Human Services, "Surgeon General Issues New Advisory About Effects Social Media Use Has on Youth Mental Health," May 23, 2023. https://www.hhs.gov/about/news/2023/05/23/surgeon-general-issues-new-advisory-about-effects-social-media-use-has-youth-mental-health.html.

7 **hours per day on social media:** Jonathan Rothwell, "Teens Spend Average of 4.8 Hours on Social Media per Day," Gallup, November 13, 2023. https://news.gallup.com/poll/512576/teens-spend-average-hours-social-media-per-day.aspx#:~:text=This%20use%20amounts%20to%204.8,for%2017%22year%2Folds.

8 **mental health harms for adolescents:** Vivek Murthy, "Surgeon General: Why I'm Calling for a Warning Label on Social Media Platforms," The New York Times, June 17, 2024, https://www.nytimes.com/2024/06/17/opinion/social-media-health-warning.html.

9 **activity of 5.4 million children:** Heather Kelly, "What Parents Need to Know About Social Media for Kids," *Washington Post*, September 30, 2021. https://www.washingtonpost.com/technology/2021/03/24/instagram-kids-faq/.

10 **A separate study:** Rachel Ehmke, "What Selfies Are Doing to Self-Esteem," Child Mind Institute. Accessed June 3, 2024. https://childmind.org/article/what-selfies-are-doing-to-girls-self-esteem/.

11 **went completely ignored:** Lauren Feiner, "Meta Failed to Act to Protect Teens, Second Whistleblower Testifies," CNBC, November 7, 2023. https://www.cnbc.com/2023/11/07/meta-failed-to-act-to-protect-teens-second-whistleblower-testifies.html.

12 **some of the biggest U.S. brands:** Jeff Horwitz and Katherine Blunt, "Instagram's Algorithm Delivers Toxic Video Mix to Adults Who Follow Children," *The Wall Street Journal*, November 27, 2023. https://www.wsj.com/tech/meta-instagram-video-algorithm-children-adult-sexual-content-72874155.

13 **underage-sex content:** Jeff Horwitz and Katherine Blunt, "Instagram Connects Vast Pedophile Network," *The Wall Street Journal*, June 7, 2023. https://www.wsj.com/articles/instagram-vast-pedophile-network-4ab7189.

14 **"Facebook Files" exposé:** "The Facebook Files: A *Wall Street Journal* Investigation," *The Wall Street Journal*, n.d. https://www.wsj.com/articles/the-facebook-files-11631713039.

15 **"quite small":** Georgia Wells, Jeff Horowitz, Deepa Seetharaman, "Facebook Knows Instagram Is Toxic for Teen Girls, Company Documents Show," *Wall Street Journal*, September 14, 2021. https://www.wsj.com/articles/facebook-knows-instagram-is-toxic-for-teen-girls-company-documents-show-11631620739.

16 **Meta-owned products:** Daniel Shvartsman, "Facebook: The Leading Social Platform of Our Times," *Investing*, April 25, 2024. https://www.investing.com/academy/statistics/facebook-meta-facts/.

17 **6 to 12 with its Messenger Kids app:** Brett Molina and Terry Collins, "Facebook Postponing Instagram for Kids Amid Uproar from Parents, Lawmakers," *USA Today*, September 27, 2021. https://www.usatoday.com/story/tech/2021/09/27/instagram-kids-version-app-children-pause/5881425001/.

18 **2022 Pew Research report:** Katherine Schaeffer, "9 Facts About Bullying in the U.S.," Pew Research Center, November 17, 2023. https://www.pewresearch.org/short-reads/2023/11/17/9-facts-about-bullying-in-the-us/#:~:text=About%20half%20of%20U.S.%20teens,it%20is%20not%20a%20problem.

19 **children in America:** Mark Weinstein, "TikTok Is a Huge Threat to Our Democracy and Our Kids. Ban It Now," *Newsweek*, December 8, 2022. https://www.newsweek.com/tiktok-huge-threat-our-democracy-our-kids-ban-it-now-opinion-1765687.

20 **not a foreign adversary:** Todd Spangler, "Senate Passes TikTok Ban Bill, Setting Up Legal Battle Between App and U.S. on First Amendment Issues," *Variety*, April 23, 2024, https://variety.com/2024/digital/news/senate-passes-tiktok-ban-bill-first-amendment-1235979220/.

21 **promoting eating disorders:** Chris Murphy, "Algorithms Are Making Kids Desperately Unhappy," *New York Times*, July 18, 2023. https://

www.nytimes.com/2023/07/18/opinion/big-tech-algorithms-kids-discovery.html.

22 **United States and the West:** Weinstein, "TikTok Is a Huge Threat to Our Democracy and Our Kids."

23 **to be a social media influencer:** Hillary Hoffower, "Kids in the US and China Have Starkly Different Goals, As Revealed by a Survey That Asked Them If They'd Rather Become Astronauts or YouTubers," *Business Insider*, July 17, 2019. https://www.businessinsider.com/american-kids-dream-of-being-youtube-influencers-instead-of-astronauts-2019-7#:~:text=American%20kids%20would%20rather%20be,millions%20of%20dollars%20a%20year.

24 **the *Washington Post* revealed:** Naomi Nix, "Meta Says Its Parental Controls Protect Kids. But Hardly Anyone Uses Them," *Washington Post*, January 30, 2024. https://www.washingtonpost.com/technology/2024/01/30/parental-controls-tiktok-instagram-use/.

25 **kids to be using social media:** Allison Gordon and Pamela Brown, "Surgeon General says 13 is 'too early' to join social media," CNN, January 29, 2023, https://www.cnn.com/2023/01/29/health/surgeon-general-social-media/index.html

26 **policy counsel at Consumer Reports:** Charlotte Morabito, "Warning labels in the U.S. seem to be everywhere. Here's why they may be pointless," CNBC, July 23, 2023, https://www.cnbc.com/2023/07/23/why-most-consumers-ignore-warning-labels.html.

27 **can increase a product's appeal:** Ziv Carmon, Yael Steinhart, and Yaacov Trope, "Scary Health Warnings Can Boost Sales," Harvard Business Review, October 2013, https://hbr.org/2013/10/scary-health-warnings-can-boost-sales.

28 **70% of U.S. adults:** Monica Anderson and Michelle Faverio, "81% of U.S. adults – versus 46% of teens – favor parental consent for minors to use social media," Pew Research Center, October 31, 2023, https://www.pewresearch.org/short-reads/2023/10/31/81-of-us-adults-versus-46-of-teens-favor-parental-consent-for-minors-to-use-social-media/.

29 **create a social media account:** Monica Anderson and Michelle Faverio, "81% of U.S. adults – versus 46% of teens – favor parental consent for minors to use social media," Pew Research Center, October 31, 2023, https://www.pewresearch.org/short-reads/2023/10/31/81-of-us-adults-versus-46-of-teens-favor-parental-consent-for-minors-to-use-social-media/.

30 **minors can spend on social media:** Monica Anderson and Michelle Faverio, "81% of U.S. adults – versus 46% of teens – favor parental consent for minors to use social media," Pew Research Center, October

31,2023,https://www.pewresearch.org/short-reads/2023/10/31/81-of-us-adults-versus-46-of-teens-favor-parental-consent-for-minors-to-use-social-media/.

31 **you know about Snapstreaks:** Snapchat, "How Do Snapstreaks Work and When Do They Expire?" help.snapchat.com/hc/en-us/articles/7012394193684-How-do-Snapstreaks-work-and-when-do-they-expire#:~:text=How%20do%20I%20keep%20my,Snapstreak%20is%20about%20to%20expire!

32 **regret his invention:** Samantha Culp, "There's an Alternative to the Infinite Scroll," WIRED, September 18, 2023. https://www.wired.com/story/lexicon-scroll-doomscrolling-mindfulness-linguistics/#:~:text=The%20dawn%20of%20the%20infinite,Instagram%20following%20in%202016)%20fully.

33 **hours of sleep per night:** Jay Summer and Nilong Vyas, "Much Sleep Should a Teenager Get?" Sleep Foundation Org, December 21, 2023. https://www.sleepfoundation.org/teens-and-sleep/how-much-sleep-does-a-teenager-need.

34 **affecting the size of teens brains:** Catherine Pearson, "How Parents Can Actually Help Teens Navigate Social Media," *New York Times*, May 23, 2023. https://www.nytimes.com/2023/05/15/well/family/kids-social-media.html.

35 **targets for commercial persuasion:** American Psychological Association, "Television Advertising Leads to Unhealthy Habits in Children; Says APA Task Force," 2004. https://www.apa.org/news/press/releases/2004/02/children-ads.

36 **as soon as preschool:** Gwen Dewar, "Teaching Critical Thinking: An Evidence-Based Guide," *Parenting Science*, 2012. https://parentingscience.com/teaching-critical-thinking/.

37 **college students and educators:** Stefanie Koperniak, "Fostering Media Literacy in the Age of Deepfakes," *MIT News*, February 17, 2022. https://news.mit.edu/2022/fostering-media-literacy-age-deepfakes-0217.

38 **struggle with that:** Pearson, "How Parents Can Actually Help Teens Navigate Social Media."

39 **K-12 students during school days:** Yi-Jin Yu, "2nd-largest school district votes to ban cellphones and social media for students," ABC News, June 18, 2024, https://abcnews.go.com/GMA/Family/2nd-largest-school-district-vote-cell-phone-social/story?id=111210584.

40 **their phones during conversations:** Heather Kelly, "Parents Have a Problem with Screen Time, Too, Teens Say," *The Washington Post*, March 11, 2024. https://www.washingtonpost.com/technology/2024/03/11/screentime-teens-parents-pew/.

Chapter 14: Surprise! Social Media Can Be Good for Your Mental Health

1 **anxiety disorder by 20%:** Dylan Walsh, "Study: Social Media Use Linked to Decline in Mental Health." MIT Sloan School of Management, September 14, 2022. https://mitsloan.mit.edu/ideas-made-to-matter/study-social-media-use-linked-to-decline-mental-health.

2 **(or positive) mental health effects:** University of Oxford, "No 'Smoking Gun' Mental Health Harm from Internet: Landmark Oxford Survey," November 28, 2023. https://www.ox.ac.uk/news/2023-11-28-no-smoking-gun-mental-health-harm-internet-landmark-oxford-survey.

3 **researchers at Harvard University:** Mesfin A. Bekalu, Rachel F. McCloud, and K. Viswanath, "Association of Social Media Use with Social Well-Being, Positive Mental Health, and Self-Rated Health: Disentangling Routine Use from Emotional Connection to Use," *Health Education & Behavior*, November 19, 2019. doi:10.1177/1090198119863768.

4 **bipolar disorder, a study:** J. A. Naslund, K. A. Aschbrenner, L. A. Marsch, and S. J. Bartels, "The Future of Mental Health Care: Peer-to-Peer Support and Social Media," *Epidemiology and Psychiatric Sciences*, April 2016. doi: 10.1017/S2045796015001067.

5 **seven hours per day on the Web:** Lindsey Leake, "17 Years of Your Adult Life May Be Spent Online. These Expert Tips May Help Curb Your Screen Time," *Fortune*, March 6, 2024. https://fortune.com/well/article/screen-time-over-lifespan/.

6 **half hours on social media:** Simon Kemp, "The Time We Spend on Social Media," Datareportal, January 31, 2024. https://datareportal.com/reports/digital-2024-deep-dive-the-time-we-spend-on-social-media#:~:text=Research%20from%20GWI%20reveals%20that,per%20day%20using%20social%20platforms.&text=On%20average%2C%20that%20means%20that,attributed%20to%20social%20media%20platforms.

7 **review by Western University:** Rea Alonzo, Junayd Hussain, Saverio Stranges, and Kelly K. Anderson, "Interplay between Social Media Use, Sleep Quality, and Mental Health in Youth: A Systematic Review," *Sleep Medicine Reviews*, April 1, 2021. doi:10.1016/j.smrv.2020.101414.

8 **published by the American Psychological Association:** Chris Palmer, "In Brief: Limiting Social Media Boosts Mental Health, the Negatives of Body Positivity, and More Research," *Monitor on Psychology*, November 1, 2023. https://www.apa.org/monitor/2023/11/benefits-limiting-social-media.

9 **select "Screen Time":** Apple, "Use Screen Time on your iPhone or iPad." https://support.apple.com/en-us/108806#:~:text=To%20see%20the%20report%2C%20go,up%20or%20received%20a%20notification.
10 **select "Digital Wellbeing":** Android, "New Ways to Find Balance for You and Your Family." https://www.android.com/digital-wellbeing/.
11 **2022 study at the University of Bath:** University of Bath, "Social Media Break Improves Mental Health, Study Suggests," *Science Daily*, May 6, 2022. https://www.sciencedaily.com/releases/2022/05/220505213404.htm.
12 **some of us get hundreds:** Molly Glick, "Phone Notifications Are Messing with Your Brain," *Discover Magazine*, April 29, 2022. https://www.discovermagazine.com/technology/phone-notifications-are-messing-with-your-brain.
13 **"Alone Together":** Sherry Turkle, *Alone Together: Why We Expect More from Technology and Less from Each Other*. Basic Books, 2011. https://www.amazon.com/Alone-Together-Expect-Technology-Other/dp/0465031463.

Chapter 15: Is AI the High-Tech Tattletale in Your Social Experience?

1 **monthly active users in in just two months:** Maria Diaz, "Threads Becomes Fastest-Growing App Ever, with 100 Million Users in Under a Week," ZDNet, July 10, 2023. https://www.zdnet.com/article/threads-hit-100-million-users-in-under-a-week-breaking-chatgpts-record/.
2 **powered by ChatGPT:** Snapchat Support, "What Is My AI on Snapchat and How Do I Use It?" https://help.snapchat.com/hc/en-us/articles/13266788358932-What-is-My-AI-on-Snapchat-and-how-do-I-use-it-.
3 **"what to make for dinner":** Ibid.
4 **even for mental health treatment:** Bernard Marr, "AI in Mental Health: Opportunities and Challenges in Developing Intelligent Digital Therapies," *Forbes*, July 6, 2023, https://www.forbes.com/sites/bernardmarr/2023/07/06/ai-in-mental-health-opportunities-and-challenges-in-developing-intelligent-digital-therapies/?sh=43b0ad325e10#:~:text=As%20we%27ve%20shown%2C%20it%27s,treatment%20plans%2C%20and%20ensuring%20compliance.
5 **"it wrote it for me":** Geoffrey A. Fowler, "Snapchat Tried to Make a Safe AI. It Chats with Me About Booze and Sex," *Washington Post*, March 14, 2023, https://www.washingtonpost.com/technology/2023/03/14/snapchat-myai/.

6 **lie to their parents about it:** Tristan Harris, "Offered advice about having sex," March 11, 2023. https://twitter.com/tristanharris/status/1634299911872348160?s=46&t=dKA6bnISNJikb2vECGEsKw.
7 **respond to their friends' texts:** Ankita Chakravarti, "Don't Like Texting? Soon ChatGPT Can Reply to Your WhatsApp Texts on Your Behalf," *India Today*, February 23, 2023. https://www.indiatoday.in/technology/news/story/dont-like-texting-soon-chatgpt-can-reply-to-your-whatsapp-texts-on-your-behalf-2338348-2023-02-23.
8 **her matches on dating apps:** Jordan Parker Erb, "I Asked ChatGPT to Reply to My Hinge Matches. No One Responded," *Business Insider*, January 11, 2023. https://www.businessinsider.com/chatgpt-replied-hinge-matches-dating-app-ai-2023-1.
9 **help you generate content:** Colleen Christison, "The 15 Best AI Tools for Social Media in 2023," Hootsuite, May 25, 2023. https://blog.hootsuite.com/best-ai-tools-for-social-media/.
10 **demonstrated this is possible:** Melissa Heikkilä, "A Watermark for Chatbots Can Expose Text Written by an AI," *MIT Technology Review*, January 27, 2023. https://www.technologyreview.com/2023/01/27/1067338/a-watermark-for-chatbots-can-spot-text-written-by-an-ai/.
11 **Instagram, and Threads:** Nick Clegg, "Labeling AI-Generated Images on Facebook, Instagram and Threads," Meta, February 6, 2023. https://about.fb.com/news/2024/02/labeling-ai-generated-images-on-facebook-instagram-and-threads/.

Chapter 16: Lifting the Veil on Bots and Trolls

1 **and interfere with democracies:** Steve Zurier, "Vast Majority of Bot Attacks Emanate from China and Russia," SC Media, September 19, 2023. https://www.scmagazine.com/news/vast-majority-of-bot-attacks-emanate-from-china-and-russia.
2 **US state department:** Dustin Volz and Michael R. Gordon, "China Is Investing Billions in Global Disinformation Campaign, U.S. Says," *Wall Street Journal*, September 28, 2023. https://www.wsj.com/world/china/china-is-investing-billions-in-global-disinformation-campaign-u-s-says-88740b85.
3 **Ukraine's military intelligence:** Daryna Antoniuk, "Ukraine Expects Billion-Dollar Russian Disinformation Campaign to Peak This Spring," *The Record*, February 28, 2024. https://therecord.media/ukraine-russia-disinformation-campaign-peaking.
4 **Musk's Twitter/X followers were fake:** Darragh Roche, "Half of Joe Biden's Twitter Followers Are Fake, Audit Reveals," *Newsweek*, May

17, 2022. https://www.newsweek.com/half-joe-biden-twitter-followers-are-fake-audit-elon-musk-1707244#:~:text=The%20software%20company%27s%20tool%20also.

5 **his Twitter/X posts were from bots:** Emily Tonelli, "Elon Musk Says 90% of Twitter Comments Are Bots—Binance CEO CZ Agrees," Decrypt, September 6, 2022. https://decrypt.co/109042/elon-musk-says-90-twitter-comments-are-bots-binance-ceo-cz-agrees.

6 **detect only 1% of their fake accounts:** Joseph Menn, "Russians Boasted That Just 1% of Fake Social Profiles Are Caught, Leak Shows," *Washington Post*, April 16, 2023. https://www.washingtonpost.com/technology/2023/04/16/russia-disinformation-discord-leaked-documents/#.

7 **started by just 1% of accounts:** Srijan Kumar, William L. Hamilton, Jure Leskovec, and Dan Jurafsky, "Community Interaction and Conflict on the Web," The Web Conference (WWW). 2018, https://snap.stanford.edu/conflict/.

8 **As reported by MIT Technology Review:** Karen Hao, "Troll Farms Reached 140 Million Americans a Month on Facebook Before 2020 Election, Internal Report Shows," *MIT Technology Review*, September 16, 2021. https://www.technologyreview.com/2021/09/16/1035851/facebook-troll-farms-report-us-2020-election/.

9 **360 million global users every week:** Ibid.

10 **protests happening in China in 2022:** Dan Milmo, "China Accused of Flooding Social Media with Spam to Crowd Out Protest News," *The Guardian*, December 4, 2022. https://www.theguardian.com/world/2022/dec/04/china-accused-of-flooding-social-media-spam-covid-protests.

11 **"defeat the bots & trolls":** Elon Musk, "I will explain the rationale," November 1, 2022. https://twitter.com/elonmusk/status/1587314744754683905?s=20&t=gZZ63rDIyYbPlBhq_nQIVQ.

12 **Twitter and democracy:** Mark Weinstein, "Elon Musk Can Save Twitter—and Democracy," *Wall Street Journal*, November 20, 2022. https://www.wsj.com/articles/elon-musk-can-save-twitter-democracy-verified-social-media-platform-revenue-content-users-features-tweet-bots-trolls-11668971538.

13 **detected for what they are:** Bill Chappell, "Deepfakes Exploiting Taylor Swift Images Exemplify a Scourge with Little Oversight," NPR, January 26, 2024. https://www.npr.org/2024/01/26/1227091070/deepfakes-taylor-swift-images-regulation.

14 **things they didn't:** Mark Weinstein, "AI Deepfakes Are Endangering Democracy. Here Are 4 Ways to Fight Back," Fox News, March 13, 2024. https://www.foxnews.com/opinion/ai-deepfakes-are-endangering-democracy-here-are-ways-fight-back.

15 **elsewhere on social media:** Dustin Volz, "China Is Targeting U.S. Voters and Taiwan with AI-Powered Disinformation," *Wall Street Journal*, April 5. 2024. https://www.wsj.com/politics/national-security/china-is-targeting-u-s-voters-and-taiwan-with-ai-powered-disinformation-34f59e21?mod=hp_lead_pos3.
16 **Georgetown and Stanford:** Chris Melore, "AI Creates Propaganda Just as Scary and Persuasive as People," StudyFinds, February 20, 2024. https://studyfinds.org/ai-creates-propaganda-persuasive/.
17 **AI expert Gary Marcus told the BBC in 2023:** David Silverberg, "Could AI Swamp Social Media with Fake Accounts?" BBC, February 13, 2023. https://www.bbc.com/news/business-64464140.

Chapter 17: Balancing Act: Free Speech versus Moderation

1 **"how they talk about politics":** Dante Chinni, "Americans More Skeptical About Social Media Than Rest of World," NBC News, January 1, 2023. https://www.nbcnews.com/meet-the-press/first-read/americans-skeptical-social-media-rest-world-rcna63045.
2 **"as well as those of the speaker":** Fredrick Douglass, "A Plea for Free Speech in Boston (1860)," National Constitution Center. https://constitutioncenter.org/the-constitution/historic-document-library/detail/frederick-douglass-a-plea-for-free-speech-in-boston-1860.
3 **refunded its investors:** Jose Pagliery, "'Secret' App Shuts Down and (Surprise!) Gives Investors Money Back," CNN Business, April 29, 2015. https://money.cnn.com/2015/04/29/technology/secret-closes-investors/index.html.
4 **viewpoints they disagree with:** Monica Anderson, "Americans' Views of Technology Companies," Pew Research, April 29, 2024. https://www.pewresearch.org/internet/2024/04/29/americans-views-of-technology-companies-2/?utm_source=AdaptiveMailer&utm_medium=email&utm_campaign=24-04-29%20Tech%20companies&org=982&lvl=100&ite=13851&lea=3344991&ctr=0&par=1&trk=a0DQm000001hMhRMAU.
5 **bear to mock leader Xi Jinping:** Javier C. Hernández, "China Censors Winnie-the-Pooh on Social Media," *New York Times*, July 17, 2017. https://www.nytimes.com/2017/07/17/world/asia/china-winnie-the-pooh-censored.html.
6 **that didn't violate its Terms of Service:** Dan Tynan, "Facebook Accused of Censorship After Hundreds of US Political Pages Purged," *The Guardian*, October 17, 2018. https://www.theguardian.com/

technology/2018/oct/16/facebook-political-activism-pages-inauthentic-behavior-censorship.

7 **when Twitter/X users clicked on their links:** Jeremy B. Merrill and Drew Harwell, "Elon Musk's X Is Throttling Traffic to Websites He Dislikes," *Washington Post*, August 16, 2023. https://www.washingtonpost.com/technology/2023/08/15/twitter-x-links-delayed/.

8 **As reported by *The Guardian*:** Alex Hern, "Revealed: How TikTok Censors Videos That Do Not Please Beijing," *The Guardian*, September 25, 2019. https://www.theguardian.com/technology/2019/sep/25/revealed-how-tiktok-censors-videos-that-do-not-please-beijing.

9 **"it's working beautifully":** David Smooke, "'The Abandonment of Clean Algos Is the Suicide of Mainstream Social Media'—Minds CEO Bill Ottman," Hackernoon, September 7, 2020. https://hackernoon.com/the-abandonment-of-clean-algos-is-the-suicide-of-mainstream-social-media-minds-ceo-bill-ottman-7o483er0.

10 **law and computer science at Harvard:** Jonathan Zittrain, "A Jury of Random People Can Do Wonders for Facebook," *The Atlantic*, November 14, 2019. https://www.theatlantic.com/ideas/archive/2019/11/let-juries-review-facebook-ads/601996/.

11 **the Wall Street Journal in 2021:** Mark Weinstein, "Small Sites Need Section 230 to Compete," *The Wall Street Journal*, January 25, 2021. https://www.wsj.com/articles/small-sites-need-section-230-to-compete-11611602173.

12 **the Los Angeles Times in 2023:** Mark Weinstein, "Op-Ed: Here's How to Reform the Law That Made the Internet," *Los Angeles Times*, February 3, 2023. https://www.latimes.com/opinion/story/2023-02-03/internet-supreme-court-case-google-section-230.

13 **comedians like Conan O'Brien:** Conan O'Brien, "I don't need to buy anything," November 23, 2018. https://twitter.com/conanobrien/status/1066081591011803137.

14 **"important law protecting internet speech":** Casey Newton, "Everything You Need to Know About Section 230," The Verge, December 30, 2020. https://www.theverge.com/21273768/section-230-explained-internet-speech-law-definition-guide-free-moderation.

Chapter 18: Facts, Opinions, Lies: Who Decides?

1 **"have no basis in fact":** Meta, "Understanding Facebook's Fact-Checking Program," November 22, 2019. https://www.facebook.com/government-nonprofits/blog/misinformation-resources#:~:text=Meta%20started%20its%20fact%2Dchecking,are%20timely%2C%20trending%20and%20consequential.

2 **grant to the IFCN:** Olivia Ma and Brandon Feldman, "How Google and YouTube Are Investing in Fact-Checking," Google News Initiative, November 29, 2022. https://blog.google/outreach-initiatives/google-news-initiative/how-google-and-youtube-are-investing-in-fact-checking/.

3 **in this space is FactCheck.org:** FactCheck.org, https://www.factcheck.org.

4 **in regard to events, science, medicine, politics, and more:** "Our Mission," FactCheck.org. Accessed on June 3, 2024. https://www.factcheck.org/about/our-mission.

5 **oldest newspaper, the *New York Post*:** Robert McMillan, "Twitter Unlocks New York Post Account After Two-Week Standoff," *The Wall Street Journal*, October 30, 2020. https://www.wsj.com/articles/twitter-reinstates-new-york-post-account-11604096659.

6 **COVID-19 originated in a lab:** Ryan Tracy, "Facebook Bowed to White House Pressure, Removed Covid Posts," *Wall Street Journal*, July 28, 2023. https://www.wsj.com/articles/facebook-bowed-to-white-house-pressure-removed-covid-posts-2df436b7.

7 **now deeming that claim to be likely true:** Hannah Rabinowitz, "FBI Director Wray Acknowledges Bureau Assessment That Covid-19 Likely Resulted from Lab Incident," *CNN*, March 1, 2023. https://www.cnn.com/2023/02/28/politics/wray-fbi-covid-origins-lab-china/index.html.

8 **removed from the internet:** Mario Cacciottolo, "The Streisand Effect: When Censorship Backfires," BBC News, June 15, 2021. https://www.bbc.com/news/uk-18458567.

9 **According to MIT research:** Peter Dizikes, "The Catch to Putting Warning Labels on Fake News," *MIT News*, March 2, 2020. https://news.mit.edu/2020/warning-labels-fake-news-trustworthy-0303.

10 **banned in certain countries:** Banned Library, "Nineteen Eighty-Four (1984) by George Orwell," January 1, 2017. https://www.bannedlibrary.com/podcast/2016/12/28/nineteen-eighty-four-1984-by-george-orwell.

11 **fact-checker at Newtral:** Lydia Morrison, "Fact-Checkers Are Scrambling to Fight Disinformation with AI," WIRED, February 1, 2023. https://www.wired.com/story/fact-checkers-ai-chatgpt-misinformation/?redirectURL=https%3A%2F%2Fwww.wired.com%2Fstory%2Ffact-checkers-ai-chatgpt-misinformation%2F.

12 **Musk later deleted his tweet:** James Felton, "Elon Musk Learned a Basic Fact About Twitter After the Platform He Owns Fact-Checked Him," *IFL Science*, November 14, 2022. https://www.iflscience.com/elon-musk-learned-a-basic-fact-about-twitter-after-the-platform-he-owns-fact-checked-him-66188.

13 **"researchers from Michigan State University revealed":** Eric W. Dolan, "Critical Thinking Education Trumps Banning and Censorship in Battle Against Disinformation, Study Suggests," PSY Post, August 9, 2023. https://www.psypost.org/2023/08/critical-thinking-education-trumps-banning-and-censorship-in-battle-against-disinformation-study-suggests-167711.
14 **"strategy for curbing disinformation":** Ibid.

Chapter 19: Seven Lessons for the Street Fight

1 **usurped by newer players:** Jeff Desjardins, "Most Valuable U.S. Companies Over 100 Years," Visual Capitalist, November 14, 2017. https://www.visualcapitalist.com/most-valuable-companies-100-years/.
2 **1 billion acres by 2050:** EarthDay Organization, "Food Doesn't Grow on Its Own." https://www.earthday.org/campaign/regenerative-agriculture/.
3 **electricity production by 2025:** International Energy Agency, "Electricity Market Report 2023," March 2023. https://iea.blob.core.windows.net/assets/255e9cba-da84-4681-8c1f-458ca1a3d9ca/ElectricityMarketReport2023.pdf.
4 **$100 million in revenue in 2023:** Jeff Beckman, "Eye-Popping DuckDuckGo Statistics 2023," TechReport, November 5, 2023. https://techreport.com/statistics/duckduckgo-statistics/.
5 **$500 million annually:** Richard Speed, "Mozilla CEO Pockets a Packet, Asks Biz to Pick Up Pace the 'Mozilla Way,'" *The Register*, January 2, 2024. https://www.theregister.com/2024/01/02/mozilla_in_2024_ai_privacy/.
6 **$260 million in 2023:** A41, "[Web3] BAT Series #3: Brave Browser's Valuation Compared to Opera and Firefox," *A41.io* (blog), July 6, 2023. https://medium.a41.io/web3-bat-series-3-brave-browsers-valuation-compared-to-opera-and-firefox-17d4faf39ea8.
7 **he didn't start it:** Andrew J. Hawkins, "How Elon Musk Took Over Tesla Using Money, Strong-arm Tactics, and His Own Popularity," The Verge, August 2, 2023. https://www.theverge.com/23815634/tesla-elon-musk-origin-founder-twitter-land-of-the-giants.
8 **love for the environment:** Patagonia, "Company History." https://www.patagonia.com/company-history/.
9 **Texas in 1980:** Whole Foods Market, "Whole Foods Market History." https://www.wholefoodsmarket.com/company-info/whole-foods-market-history.

10 **"REI member dividend":** Kristen Bor, "12 Reasons to Become an REI Member," *Bearfoot Theory* (blog), March 26, 2023. https://bearfoottheory.com/reasons-to-become-rei-member-benefits/#:~:text=Every%20year%20in%20March%2C%20REI,or%20via%20the%20REI%20app.
11 **accountability, and transparency:** B Corporation, "Measuring a Company's Entire Social and Environmental Impact," October 6, 2023. https://www.bcorporation.net/en-us/certification/.
12 **effective regulation to this day:** US Environmental Protection Agency, "Superfund." https://www.epa.gov/superfund.
13 **if maximally enforced:** Craig Timberg, "How Big Could Facebook's Fine Theoretically Get? Here's a Hint: There Are Four Commas, and Counting," *Washington Post*, April 9, 2018. https://www.washingtonpost.com/news/the-switch/wp/2018/04/09/how-big-could-facebooks-fine-theoretically-get-heres-a-hint-there-are-four-commas-and-counting/?noredirect=on.
14 **explicitly promising not to:** BBC, "Google to Pay Record $391m Privacy Settlement," November 15, 2022. https://www.bbc.com/news/technology-63635380.
15 **in annual advertising revenue:** Ibid.
16 **used the data for targeted advertising:** Cat Zakrzewski, "Twitter Will Pay $150 Million Fine over Deceptively Collected Data," *The Washington Post*, May 25, 2022. https://www.washingtonpost.com/technology/2022/05/25/twitter-fine-ftc/.
17 **well-resourced companies in the world:** Ibid.
18 **10% of a company's gross revenue:** Katie Collins, "Big Tech Behind Bars? The UK's Online Safety Bill Explained," CNET, January 19, 2023. https://www.cnet.com/news/politics/big-tech-behind-bars-the-uks-online-safety-bill-explained/.

Chapter 20: Welcome to Web4: The Restoration of Sanity

1 **European Commission:** James Cirrone, "What Is Web4? The EU Wants to Explain," Blockworks, July 11, 2023. https://blockworks.co/news/eu-wants-web4.
2 **oil and gas industry's 23%:** Chantal d'Offay, "Trust in Social Media," *IPSOS*, March 30, 2023. https://www.ipsos.com/en/trust/trust-social-media#:~:text=It%20is%20in%20this%20context,vs%2016%25%2C%202021.
3 **respond to suspicious behavior:** "Cybersecurity and ChatGPT: Use Bots to Fight Bots," *RSA Blog*, January 23, 2023. https://www.rsa.com/

products-and-solutions/cybersecurity-and-chatgpt-use-bots-to-fight-bots/.

4 **concerned about their privacy online:** Müge Fazlioglu, "Most Consumers Want Data Privacy and Will Act to Defend It," IAPP, March 22, 2023. https://iapp.org/news/a/most-consumers-want-data-privacy-and-will-act-to-defend-it/.

5 **trade-off for "free services:"** Mark Murray, "Poll: Americans Give Social Media a Clear Thumbs-Down," NBC News, April 5, 2019. https://www.nbcnews.com/politics/meet-the-press/poll-americans-give-social-media-clear-thumbs-down-n991086?curator=TechREDEF.

6 **according to an Axios survey:** Kyle Daly, "Exclusive: Poll Reveals Americans' Data Privacy Frustrations," Axios, August 13, 2020. https://www.axios.com/2020/08/13/exclusive-poll-reveals-americans-data-privacy-frustrations.

7 **the *Los Angeles Times*:** Mark Weinstein, "Opinion: Americans Might Finally Get a Real Privacy Law to Fight Big Tech Intrusions," *Los Angeles Times*, April 29, 2024. https://www.latimes.com/opinion/story/2024-04-29/americans-privacy-law-big-tech.

8 **because it drives higher engagement:** David Streitfeld, "'The Internet Is Broken': @ev Is Trying to Salvage It," *New York Times*, May 20, 2017. https://www.nytimes.com/2017/05/20/technology/evan-williams-medium-twitter-internet.html.

9 **address environmental and social issues:** "Patagonia Launches Venture Capital Fund," Sustainable Business, https://www.sustainablebusiness.com/2013/05/patagonia-launches-venture-capital-fund-51519/#:~:text=Patagonia%2C%20one%20of%20the%20most,water%2C%20energy%2C%20and%20waste.

10 **working on sustainable products and services:** Unilever, "Unilever Launches Global Platform to Engage with Start-ups," March 22, 2014. https://www.prnewswire.com/news-releases/unilever-launches-global-platform-to-engage-with-start-ups-260251991.html.

11 **cold-pressed juice company Suja:** John Kell, "Coke Invests in Fast-Growing Organic Juice Maker Suja," *Fortune*, August 20, 2015. https://fortune.com/2015/08/19/coca-cola-investment-suju-juice/.

Chapter 21: Taking It to the Streets

1 **elephant during a vacation to India:** Jan Crawford, "Opposites Attract: A Look Inside the Unlikely Friendship of Scalia and Ginsburg," CBS News, February 15, 2016, https://www.cbsnews.com/news/opposites-attract-a-look-inside-the-friendship-of-antonin-scalia-and-ruth-bader-ginsburg/.

APPENDIX A: Advice to Legislators

1. **perpetrate upon kids:** Cecilia Wang, "'Your Product Is Killing People': Tech Leaders Denounced Over Child Safety," *New York Times*, January 31, 2023, https://www.nytimes.com/2024/01/31/technology/senate-child-safety-social-media.html.
2. **"what materially has occurred?":** Amanda Silberling, "Senate Hearing with Five Social Media CEOs Was a Missed Opportunity," TechCrunch, January 31, 2024. https://techcrunch.com/2024/01/31/senate-hearing-with-five-social-media-ceos-was-a-missed-opportunity/.
3. **issued a dire public advisory:** US Department of Health and Human Services, "Social Media and Youth Mental Health: The U.S. Surgeon General's Advisory: 2023," https://www.hhs.gov/sites/default/files/sg-youth-mental-health-social-media-advisory.pdf.
4. **collecting information on kids:** Federal Trade Commission, "Children's Online Privacy Protection Rule ('COPPA')." https://www.ftc.gov/legal-library/browse/rules/childrens-online-privacy-protection-rule-coppa.
5. **adhere to regulations:** Sam Schechner, "U.K. to Toughen New Social-Media Law, Threatening CEOs with Jail Time," *Wall Street Journal*, January 17, 2023. https://www.wsj.com/articles/u-k-to-toughen-new-social-media-law-threatening-ceos-with-jail-time-11673963019.
6. **choose to take down:** Mark Weinstein, "The Supreme Court Could Soon Change The Internet Forever—Here's What You Need to Know," The Hill, October 24, 2023. https://thehill.com/opinion/technology/4271001-the-supreme-court-could-soon-change-the-internet-forever-heres-what-you-need-to-know/.
7. **"free exchange of ideas than to encourage it":** Jeff Kosseff, "The Internet Speech Case That the Supreme Court Can't Dodge," WIRED, August 7, 2023. https://www.wired.com/story/tech-policy-netchoice-scotus/#:~:text="As%20a%20matter%20of%20constitutional,it%2C" %20the%20Court%20wrote.
8. **"I Compete With Facebook, and It's No Monopoly":** Mark Weinstein, "I Compete with Facebook, and It's No Monopoly," *The Wall Street Journal*, June 27, 2019. https://www.wsj.com/articles/i-compete-with-facebook-and-its-no-monopoly-11561676776?mod=article_inline.
9. **competitors in personal social networking:** Brent Kendall, "Facebook Hit with New Antitrust Suit from Federal Trade Commission," *The Wall Street Journal*, August 19, 2021. https://www.wsj.com/articles/facebook-hit-with-renewed-antitrust-lawsuit-as-ftc-tries-again-11629387483?mod=article_inline.

10 **calling the FTC's lawsuit "meritless.":** Meta Newsroom, "It is unfortunate," X, August 20, 2021. https://twitter.com/fbnewsroom/status/1428436569267056648?s=2.
11 **"I Changed My Mind—Facebook Is a Monopoly":** Mark Weinstein, "I Changed My Mind—Facebook Is a Monopoly," *The Wall Street Journal*, October 1, 2021. https://www.wsj.com/articles/facebook-is-monopoly-metaverse-users-advertising-platforms-competition-mewe-big-tech-11633104247.
12 **merge together again over the years:** Jose Pagliery, "How AT&T Got Busted Up and Pieced Back Together," CNN Business, May 20, 2014. https://money.cnn.com/2014/05/20/technology/att-merger-history/index.html#:~:text=To%20tear%20down%20a%20nationwide,(T)%20we%20know%20today.
13 **it was restored nationally:** Cecilia Kang, "F.C.C. Votes to Restore Net Neutrality Rules," *New York Times*, April 25, 2024. https://www.nytimes.com/2024/04/25/technology/fcc-net-neutrality-open-internet.html.
14 **"Before It's Too Late":** Mark Weinstein, "Three Ways to Regulate AI Right Now Before It's Too Late," Fox News, April 5, 2023. https://www.foxnews.com/opinion/three-ways-regulate-ai-right-now-before-too-late.
15 **detect if content was created by AI:** Melissa Heikkilä, "A Watermark for Chatbots Can Expose Text Written by an AI," *MIT Technology Review*, January 27, 2023. https://www.technologyreview.com/2023/01/27/1067338/a-watermark-for-chatbots-can-spot-text-written-by-an-ai/.
16 **"AI Deepfakes Are Endangering Democracy":** Mark Weinstein, "AI Deepfakes Are Endangering Democracy. Here Are 4 Ways to Fight Back," Fox News, March 13, 2024. https://www.foxnews.com/opinion/ai-deepfakes-are-endangering-democracy-here-are-ways-fight-back.
17 **"Trustworthy Artificial Intelligence":** The White House, "Fact Sheet: President Biden Issues Executive Order on Safe, Secure, and Trustworthy Artificial Intelligence," October 30, 2023. https://www.whitehouse.gov/briefing-room/statements-releases/2023/10/30/fact-sheet-president-biden-issues-executive-order-on-safe-secure-and-trustworthy-artificial-intelligence/.

APPENDIX B: Accountability Structures for Restoration Networking

1 **regenerative organic certified:** Rodale Institute, "Regenerative Organic Certified." Accessed June 3, 2024. https://rodaleinstitute.org/regenerative-organic-certification/.

2 **with the Green-e program:** Green-e, "Green-e Energy." Accessed June 3, 2024. https://www.green-e.org/programs/energy.

APPENDIX C: Revenue Superchargers

1 **provide enhanced features:** Jie Yee Ong, "Music Streaming Services Explained: How Does Spotify Make Money?" MUO: Make Use Of, January 3, 2021. https://www.makeuseof.com/music-streaming-services-explained-how-does-spotify-make-money/.

2 **expended worldwide on digital advertising:** "Digital Advertising Spending Worldwide from 2021 to 2026," Statista. https://www.statista.com/statistics/237974/online-advertising-spending-worldwide/.

3 **platforms like Substack:** "Great Writing Is Valuable," Substack. Accessed June 3, 2024. https://substack.com/going-paid.

4 **average donation is approximately $15:** Debarghya Sanyal, "Why Does Wikipedia Need Donations Despite Its Massive Popularity?" *Business Standard*, October 4, 2023. https://www.business-standard.com/technology/tech-news/why-does-wikipedia-need-donations-despite-its-massive-popularity-123100400249_1.html.

Acknowledgments

It all starts with family. The very idea that I would endeavor to invent one of the first "social media" websites came from my family. Thank you, sisters, brother, nieces, and nephews! It's been a remarkable journey ever since.

The book in your hands has seen its share of editorial fingerprints. Many thanks to all of those who shepherded the text. Yet the coherence and readability of this book are keenly due to David Westreich, who has toiled beside me for the past two years with a never-ending flow of ideas, an unbridled well of enthusiasm, and an unwavering eye for sharpening the written word.

This book made the rounds in pitches with a variety of publishers. But it was Victoria Savanh at Wiley who deserves all the credit for immediately understanding its importance for the world. Victoria was steadfast in her determination to bring the book into the Wiley family and see it published just as I intended for it.

I am immensely thankful to the media, to all the editors and reporters worldwide from countless publications, who found and continue to find what I am doing and what I have to say compelling and important. Thank you for boldly interviewing and publishing me.

What is particularly heartening is that my commonsense perspective finds homes on all sides of the political and thought leader spectrum.

A great many brilliant minds have generously given their ideas, advice, and support to me. From the Web's inventor, Sir Tim Berners-Lee; to the cofounder of Apple, Steve Wozniak (Woz); to brilliant scholars like Sherry Turkle; and countless others, there are far too many to list here. My path glitters with the lucky crossings of remarkable people in our shared journey here on earth. I would be remiss if I avoided acknowledging the tough love I've received from friends and colleagues as well. Life is perfect in its imperfection, and I am utterly human. Thank you to all who criticize and challenge me; we're all works in progress, and you help me to improve and grow, every day.

Over the years, countless team members, the proverbial front lines of it all, have joined with me to bring my social media ruminations to life—software engineers, designers, dev ops, and so many other competencies that bring it all together. Disruption isn't always easy. Funding was a cycle of feast or famine, yet the abundance of funds or lack thereof never deterred your fierce loyalty and hard work.

My journey hasn't been a walk in the park with venture capitalists. Particularly when it came to upending the Meta monopoly. Silicon Valley VCs can be a bit myopic. Meta isn't kind to competitors. Instead, my companies were financed by well-to-do everyday folks and family offices who loved and supported the vision of social media done right. Thank you for your steadfast partnership in this remarkable mission to do the right thing for ourselves, our kids, democracy, and the world.

My adventure in social media has lasted more than 25 years through three award-winning personal social media platforms. There are millions of you worldwide who have accompanied me. I owe my entire learning curve to all of you. Social media seemed like a fun idea in the 1990s. But who knew that it would dominate the world in the 21st century! Thank you for your curiosity, support, and ethics—you all wanted something better than what was offered by Big Tech. The lessons we've learned together allow us to catapult now.

Finally, I thank you, the book's readers. It is your excitement that brings the book's ideas to life. It's time to get this right for humanity. Together, we're on it!

About the Author

Mark Weinstein is a world-renowned tech entrepreneur, contemporary thought leader, privacy expert, and one of the visionary inventors of social networking. From Web1 in the 1990s into Web2 in the 2020s, he's led award-winning social media companies that revolutionized the industry.

In 1998, Mark launched SuperFamily and SuperFriends, two of the earliest social networks. They were listed on PC Magazine's "Top 100 Sites" for three years. In 2016, Mark founded MeWe, the Facebook alternative with the industry's first Privacy Bill of Rights. He led MeWe as its CEO until March 2021, winning accolades including being named a "2020 Most Innovative Social Media Company" by *Fast Company* and a "2019 Best Entrepreneurial Company in America" by *Entrepreneur Magazine*. MeWe grew to nearly 20 million users worldwide with millions of dollars in revenue—an unprecedented feat with no VC funding or marketing budget.Mark brought luminaries to MeWe's Advisory Board, including Sir Tim Berners-Lee, the inventor of the Web; Steve "Woz" Wozniak, co-founder of Apple; Sherry Turkle, MIT academic and tech ethics leader; and Raj Sisodia, co-founder of the Conscious Capitalism movement. He left MeWe to work on Restoring Our Sanity Online.

A leading privacy advocate, his landmark 2020 TED Talk, "The Rise of Surveillance Capitalism," exposed the infractions and manipulations of Big Tech, and called for a privacy revolution. Mark has also been listed as one of the "Top 8 Minds in Online Privacy" and named "Privacy by Design Ambassador" by the Canadian government.

Mark is frequently interviewed and published in top-tier media including the *Wall Street Journal*, *New York Times*, Fox, CNN, BBC, *Newsweek*, *Los Angeles Times*, *The Hill*, and more. He covers topics including social media, privacy, AI, free speech, antitrust, and protecting kids online. His previous books on habits and personal greatness won two Indie Book Awards and were endorsed by Stephen Covey. Mark received his BA from the University of California, Santa Cruz; and MBA from the Anderson School of Management at UCLA.

Index

B